WORKING THE HOMESTAKE

Joseph H. Cash

WORKING THE

HOMESTAKE

JOSEPH H. CASH

THE IOWA STATE UNIVERSITY PRESS / AMES

TO MY MOTHER

JOSEPH H. CASH, a native of South Dakota, has great interests in the history of his state and in Indians. His writings include two books about Indians. He earned his B.A. and M.A. degrees at the University of South Dakota and his Ph.D. degree at the University of Iowa. He is presently Duke Research Professor of History at the University of South Dakota.

TITLE PAGE: The great mining center of the Black Hills. Lead City, Dakota Territory. 1888.

© 1973 The Iowa State University Press
Ames, Iowa 50010. All rights reserved

Composed and printed by The Iowa State University Press

First edition, 1973

Library of Congress Cataloging in Publication Data
Cash, Joseph H.
    Working the Homestake.
    Bibliography: p.
    1. Gold miners—South Dakota. 2. Homestake Mining Company. I. Title.
HD8039.M74U53   338.2'7'4109783   72-2048
ISBN 0-8138-0755-7

CONTENTS

ACKNOWLEDGMENTS

ANY WORK of research is dependent on many people for its virtues and on one for its failures. Certainly this is the case here. Thanks and gratitude in full measure are given those who did so much. By the same token, responsibility for the flaws rests solely upon me.

Many fine scholars have aided this research. Malcolm J. Rohrbough of the State University of Iowa, the supervisor of this study, has helped with his penetrating comments, high knowledge, and fine sense of justice. Allan G. Bogue of the University of Wisconsin greatly aided the early phases. Herbert S. Schell and the late Everett Sterling of the University of South Dakota started my research in Black Hills history and opened many doors to me.

The men of the Homestake Mining Company gave valuable assistance. Kenneth Keller, vice-president and chief counsel, opened the Homestake records for the first time. Donald P. Howe, the publicity director, was always most helpful, as was John Moody, the director of employment. James Harder, the present superintendent, and his assistant, Clarence Kravig, gave freely of their help; so did Dr. Arthur Semones, the chief surgeon, and Joe Dunmire, the recreation director. All illustrations used in this book were supplied through the courtesy of the Homestake Mining Company.

Three libraries have given the greatest service possible in finding materials. To the librarians and their staffs of the University of South Dakota and the State University of Iowa goes the deepest appreciation. I should like to express particular gratitude to Miss Dorette Darling and her staff of the Homestake Library in Lead, South Dakota.

Mr. and Mrs. Darrell Booth have been loyal and efficient manuscript proofreaders. It is indeed remarkable that a president of a nationwide corporation would spare his time for such work.

To my neighbors and friends in Lead who helped me in so many ways I give thanks. Some are mentioned herein, but more are not. All were gracious and gave unstintingly of their knowledge, experience, and time.

Finally, it is proper that I thank my family. My children, Sheridan, Joe, and Meredith, have lost a father and gained a book—not a good bargain for them. My mother, Mrs. J. R. Cash, has given freely of her aid. Margaret Cash, my wife, has worked harder than anyone should on this manuscript and still maintained her considerable skills in the housewifely arts. As much as anyone, she is responsible for the good elements in the finished product.

INTRODUCTION

THE PRECIOUS METAL mining industry in the American West has been approached in myriad ways by a variety of people. The gaudy days of the gold rushes have appealed to interpreters ranging from Mark Twain to television script-writers. Solid scholars such as Rodman Paul and William S. Greever have studied the gold rushes and the mining industry as a whole. In nearly all cases, however, this mining industry is regarded as a nineteenth-century phenomenon that ceased to have importance in the new century. This is not accurate, of course. Western precious metal mining continues, although it has declined considerably.

Even more neglected than twentieth-century mining in the West has been the study of labor in these mines. Vernon Jensen has contributed an excellent study of the Western Federation of Miners and Paul Brissenden a study of the Industrial Workers of the World. These, however, are institutional studies that revolve around the problems of an organization and are, by and large, approached from the viewpoint of the organization. Practically nothing has been done on the relationship of the worker to his company—his type of work, supervision, housing, recreation, and general well-being.

It would be presumptuous to suggest that the many problems raised in a study of western labor and gold mining will be solved here. Indeed, no attempt will be made to do so. Instead, the lot of the working man in one mine—the Homestake of Lead, South Dakota—will be examined.

The Homestake Mine is not a typical gold mine, and the experiences that it has undergone should not be taken as a microcosm of the industry as a whole. In fact, it is this unique quality that attracts attention. The mine has run almost continuously from 1877 to the present. It is now the largest mine in the Western Hemisphere and the only large gold mining operation remaining in the United States. This preeminence is due to a variety of factors: adequate capitalization, sound management, an enormous ore body, a willingness to adopt new devices and techniques, and a relationship with its working force that generally has been peaceful and highly productive to the interests of both company and labor.

The good relationship between management and labor came about through a program of paternalism that began with individual philanthropy and was later systematized following a lockout that ended organized labor in the company works. It has been the fashion in recent years to damn paternalism. Peter R. Drucker has said:

> The failure of paternalism is obvious. Except for a few survivals which are just barely kept alive by respect for the "Old Man" who built the business, paternalism has become as good as extinct. The reasons for its failure are obvious too. It has been proved not only a false answer but a false answer to the wrong problem. It rests on the basic fallacy that people will take propaganda for reality. Paternalism attempts to give the individual in industrial society status and function by telling him that he has status and function. The problem of status and function in industrial society arises because in the modern plant the worker does not have the dignity and responsibility of an adult but is kept in the dependence of a child. Paternalism tries to make him feel like an adult by treating him like a good child. The result has been—at least in this country—that paternalist management has often led to greater dissatisfaction than the rule of a "tough boss."
>
> Management has large responsibilities for the worker which it cannot shirk. But the solution of the problem of function and status in the industrial system cannot be found in doing more for the worker, in giving him more social security, more welfare and recreational agencies, in looking after him better. It can only lie in giving him the responsibility and dignity of an adult.[1]

Apparently the Homestake was able to provide for the security and welfare of its workers and in so doing reached a degree of labor peace and stability unusual in an industry noted for the exploitation of both human and mineral resources.

While company policy and its pervasive influences are of

great importance, the basic story concerns the workers—individually and collectively. Where they came from, why they came, what they did, how they lived, and how they worked are basic to the study of mining in the West. The corporation, with all its skills and resources, was helpless without them. They dug the gold.

WORKING THE HOMESTAKE

CHAPTER ONE

GOLD RUSH IN THE BLACK HILLS

THE PLAINS of South Dakota start at the Missouri River and rise imperceptibly toward the west. They have an abstract beauty of their own, invisible to the eye schooled to the lush greenery of areas further east. On these plains grows the grass that fed the great buffalo herds, which in turn were the basis of life for the Indians—Kiowa and Crow, Cheyenne and Sioux. By the middle of the nineteenth century the latter tribes had conquered this area and that to the west of it, now Wyoming and Montana.

Rising abruptly out of the plains are the Black Hills, the Pa-ha-sa-pa of the Sioux. These mountains, which appear black when viewed from a distance, are in the greatest possible contrast to the surrounding plains. The Sioux considered them holy and the center of their entire world. They seldom lived among the spruce and pine forests, the granite outcroppings, the high peaks, and the clean mountain streams but visited them for religious ceremonials or to cut lodge poles. No other place had such an emotional pull on the Indian.

Within the Black Hills, in the streams and the quartz-veined mountains, were the flakes of gold that would lure the white man to the mountains and eventually break the hegemony of the Sioux nation. Probably an Indian was the first to discover gold in the streams of the Black Hills.[1] It is impossible to fix a definite time or place for the discovery or even to prove positively that it took place, but its likelihood cannot be overlooked. The reaction of this unknown red man to the presence of gold can never be known, but it is improbable that he conceived of its value.

White men, however, knew the value of gold, how to find it, and how to extract it from the streams and the quartz that hid it. American technology in gold mining had advanced considerably since the Carolina discoveries in the 1790s.[2] The California Gold Rush of 1849 provided not only the basic techniques but also a large number of men who continually sought the precious metal all over the American West. As they probed, it was perhaps inevitable that they would find the gold in the Black Hills. At any rate, it is very likely that the area had been prospected intermittently for a considerable number of years prior to the actual gold rush.[3]

The early accounts of the discovery of gold in the Black Hills are unsubstantiated for the most part and impossible to authenticate. One of the earliest tells of Indians in 1811 bringing fine nuggets of gold to the trading post located at the confluence of the forks of the Cheyenne River.[4] The location would indicate this was Black Hills gold.

A tangible piece of evidence pointing toward an early discovery of gold came in 1887 when Lewis Thoen, a farmer near Spearfish, South Dakota, discovered a chunk of sandstone on Lookout Mountain with a message carved upon it:

> Came to these Hills in 1833, Seven of us; De la Compt, Ezra Kind, C. W. Wood, T. Brown, B. Kent, William King, Indian Crow. All dead but me, Ezra Kind. Killed by Indians beyond the high hill. Got our gold June, 1834. Got all of the gold we could carry our ponys all got by the Indians I have lost my gun and nothing to eat and Indians hunting me.[5]

The authenticity of the Thoen stone has never been questioned, and there is no reason to do so now. Ezra Kind has never been heard of otherwise; the Indians or starvation must have killed him. In the following year Captain Bonneville of the United States Army, passing through on an exploring expedition, discovered gold in what may have been the Black Hills and reported the find.[6]

It is highly probable that many private expeditions came to the Hills seeking for gold in an early period. There are many instances of miners during the gold rush finding evidence of earlier miners. Mining implements, gads, hatchets, and a petrified pair of pants were uncovered within the city limits of Deadwood at a depth that precluded their being abandoned near the time of discovery.[7] Old tunnels and shafts were discovered at various points in the Black Hills—in Gold Run Canyon, for example.[8] There is no way of absolutely discerning who left the implements and dug the tunnels, but it must have been white men. No Plains Indian would have possessed the material, skill, or desire to accomplish any of this.

There can be little doubt that the noted Jesuit missionary, Father DeSmet, knew of the existence of gold in the Black Hills at an early date. The good priest made a journey to the mountains in 1848 and was in close attendance to the Sioux thereafter.[9] Seth Bullock, Deadwood's first sheriff and the boon companion of Theodore Roosevelt, stated that shortly after the close of the Civil War, Father DeSmet revealed the secret of the gold at a dinner party in Columbus, Ohio.[10] Apparently he had warned the Sioux to guard their secret with care in order to prevent the rapaciousness of the white man from destroying the Indian while seeking his treasure.[11]

In 1852 the David Dale Owen Geological Survey ran a section of the Black Hills but apparently found no gold.[12] During the same year a starving man wandered into a Mormon camp on the Green River in Wyoming. His name was Thomas Renshaw, and he told a strange story. He had been one of a party that had left Council Bluffs, Iowa, for the California gold fields in 1852. They heard tales of gold in the Black Hills while at Fort Laramie, and nineteen of the party had gone in search of it. They found prospects on what may have been Rapid Creek and then went north to the high hills on Whitewood Creek where they struck bedrock and sank a shaft. Three of the party left for Fort Laramie to report to the main party. Renshaw remained and, after hunting deer one day, returned to camp and saw Indians dancing around the bodies of his companions. He left the scene, wandered for weeks, and finally made it to an immigrant road and thence to the Mormon camp.[13] This story received corroboration in 1878 when two French hunters named LeFevre came to Lead City with a tale of having discovered two skeletons. Both were men who had carried Kentucky long rifles and a leather-covered memorandum book bearing the date "1-52." It was assumed that these two had been members of the Renshaw party and that they had sunk the shaft in Rutabaga Gulch, near which they had died.[14]

In 1857 Lieutenant G. K. Warren of the United States Topographical Engineers, a man who would later become a major general and save the day at Gettysburg, visited the Black

Hills. His report on this journey states that the formations of stratified rock in the Black Hills were basically the same as those found in gold-bearing formations of the Wind River and the Big Horn Mountains.[15] Dr. F. V. Hayden, the eminent scientist accompanying Warren, said he had found considerable gold while searching for fossils.[16]

Rumors of gold in the Black Hills were reaching a highly interested public by the 1860s. The young frontier communities of southeastern Dakota were intrigued. The citizens of Yankton, the territorial capital, organized the Black Hills Exploring and Mining Association in January 1861. For a variety of reasons, such as the Civil War and the Santee Sioux uprising, this group was unable to start an expedition to the Hills until 1866–67, and this was stopped almost immediately by the United States government.[17] On March 17, 1862, Governor Jayne of Dakota Territory addressed his legislature and spoke of the mineral wealth of the Black Hills. Governor Newton Edmunds reiterated the theme in 1864, and thereafter every governor referred to the gold deposits in the Black Hills.[18] In 1867 the Black Hills Exploring and Mining Association, undaunted by its wintertime failure, revived its plan for entering the Hills. However, the government again foiled them. In 1868 the United States signed the Treaty of Fort Laramie with the Sioux. The treaty created an enormous reservation that included all of South Dakota west of the Missouri River. This meant that the Black Hills, with its well-known mineral wealth, belonged to the Indian for his exclusive and perpetual use. By the solemn word of the United States government, no white man could enter or alienate any of this territory.[19] This treaty may have postponed the entry of the Hills for a few years, but it certainly did not prevent it. When authoritative information confirming the presence of paying quantities of gold in the Black Hills combined with the economic spur of a great depression, neither army nor Indian could halt the resultant rush of miners.

There were men, however, who needed no more information and who found their motivation elsewhere. Among these was a Sioux City newspaperman, Charles Collins. In the fall of 1869 he presented a most unusual scheme to the assembled Hibernians present at the Fenian Convention in St. Louis.[20] Collins proposed that an Irish-American colony be established in the western part of Dakota Territory. Each settler would receive 160 acres of land. When the number of settlers was large enough, when enough wealth had been accumulated, and when Great Britain was in considerable trouble elsewhere, the Irish colonists would invade and conquer Canada. Having acquired Canada with very little bother, the Fenians would then arrange an exchange with Queen Victoria—all of Canada for the independence of Ireland.[21]

Surprisingly enough, this fantastic idea achieved considerable support. Congress passed a bill authorizing a colony corporation with A. T. Steward, Jim Fisk, Jr., Ben Butler, Wendell Phillips, and several other famous Americans as officers.[22] Fortunately, rationality returned before anyone got hurt. The Fenians sent an investigating group to the area which concluded that the danger from the Indians was too great.[23] This unfortunate development did not deter Charles Collins in the least. He concluded that the settlement of the Black Hills would have to be undertaken by other than the sons of Erin. He and Captain T. H. Russell launched a concerted propaganda campaign designed to lure people to the Hills.[24] This would bear some fruit a bit later.

Others with less exalted aims than Collins had ideas of getting into the Hills. In 1870 a group from Bozeman, Montana, organized a company to explore the region. The proposed expedition was to have an artillery battery, a surgeon, and press correspondents. The United States Army ordered a halt to all such expeditions in 1872, but it did not permanently deter the Montanans. They began their journey in 1874 but were forced back by the Sioux.[25]

In 1873 the United States suddenly found itself in the midst of the so-called Grant Depression. This was economic dislocation of the first magnitude. Men found themselves ruined, and multitudes became unemployed.[26] Desperation drove many to try to reestablish themselves in the new gold regions.[27]

In 1874 the federal government, goaded by the panic and by persistent rumors of gold, sent Brevet Major General George A. Custer into the Black Hills to survey the economic possibilities of the region.[28] Custer, who preferred to do things in a big way, left Fort Abraham Lincoln with ten companies of the Seventh Cavalry, one each of the Twentieth and the Seventeenth Infantry, nearly 100 Indian scouts, an engineer, a naturalist (George Bird Grinnell), a botanist, two practical gold miners, and assorted newspapermen. To transport and feed this horde came a wagon train of 110 wagons, 1,000 cavalry horses, and 300 beef cattle. He even took artillery—three gattling guns and a three-inch rifle.[29] Needless to say, the Sioux—intelligent to a man—were conspicuous by their absence.

According to the correspondent of the *New York Tribune*, the first gold was found July 30, 1874, on the site of the present city of Custer.[30] This show of color was of insufficient quantity, and on August 1, 1874, Custer moved to what would be the site of the Gordon Stockade; it was there that Horatio Nelson Ross, a practical miner, made the significant strike. Custer's favorite scout, "Lonesome Charley" Reynolds, sped with the news to Fort Laramie and the telegraph.[31]

Around the campfire that night, the first mining company organized in the Hills, the Custer Peak Mining Company, was formed. It claimed 4,000 feet, and Ross got the discovery claim of 400 feet. The miners, at Custer's suggestion, called the valley of French Creek "Golden Valley."[32]

The news of Custer's discovery struck a country that needed hope. Many men still suffering from the panic planned to go to the Hills and try their luck, danger and hardship notwithstanding.[33] General Philip Sheridan, whose command included the Black Hills and surrounding area, issued a drastic though ineffective order designed to curb trespassers in the Indian country.[34]

By December 28, 1874, the first sizable party of miners had slipped through the army defenses and reached French Creek. This was the Gordon-Russell party of Sioux City. One of its organizers was the familiar Charles Collins, and among the party was Annie D. Tallent, the first white woman known to have entered the Black Hills. The party naturally went to the place where Custer had found color; there, on February 23, 1875, they held a meeting in the stockade they had built and drafted their mining laws. They called the district the Custer Mining District and named Angus MacDonald as their recorder.[35] In so doing, they performed an act done countless times before and repeated thereafter. It was an expediency used to guard property rights and to provide a criminal code in areas that the law had not penetrated. In this case the law could not penetrate because the land belonged to the Indians, and federal law had no force whatsoever. The Gordon party had little time to enjoy their district. On April 7, 1875, the Army expelled them from the Hills, one jump ahead of the Indians. Hopeful to the end, they recorded their last claim on the day of their expulsion.[36]

Although the Gordon party left, more miners filtered through the army patrols; a few were turned back, but many made it through. A meeting held June 11, 1875, established the Cheyenne Mining District on French Creek. The claims started at the Gordon Stockade and extended to the head of the creek.[37]

The miners drifting across the plains and through the Hills were apparently convinced that there was gold, but the government was still in doubt. In an attempt to resolve this, W. P. Jenney and Henry Newton were sent into the region under strong military escort. In May 1875 they investigated the southern Hills thoroughly.[38] Jenney informed Washington that he had found extremely valuable deposits of placer gold in numerous locations.[39] The government could be certain now of the truth of the reports of gold, but it could have been no source of joy to the harassed servants of the Republic. The land still belonged to the Sioux; the miners were trespassing on it; the United States had pledged its word in perpetuity to maintain the Indian rights. Trouble was inevitable. Washington could only decide when, where, and with whom the trouble would occur.

The Grant administration, which always inclined toward a fair Indian policy in an ineffectual way, tried hard to resolve the quandary. In 1875 it attempted to negotiate a treaty of sale with the Indians. This failed for a variety of reasons, among them the fact that Charles Collins was deliberately sabotaging the negotiations in a manner that nearly got all parties massacred.[40] With the failure, the government called off the army, and the miners swarmed into the Hills.[41] An indirect result of these failures was the Sioux war that resulted in the massacre of Custer's command on the Little Big Horn in 1876. Grant, in his message to Congress in 1876–77, explained the reasons for withdrawal of the troops:

Hostilities there have grown out of the avarice of the white man, who had violated our treaty stipulations in his search for gold. . . . Gold had actually been found in paying quantities, and our efforts to remove the miners would only result in the desertion of the troops that might be sent there to remove them.[42]

Grant's explanation was that of an extremely pragmatic soldier and is quite convincing. The government obviously could not keep the miners out of the Hills; any further attempt to do so would result in weakening the army to no purpose.

The miners continued to pour into the Hills. A Black Hills veteran saw three distinct types of men in the rush. Unsettled Civil War veterans, still footloose some ten years after the end of the war, composed the first. The second type included men who had fallen before the onslaught of the panic of 1873. The third encompassed professional miners from the Montana, Nevada, California, and Colorado gold camps, with an added dash provided by the flotsam of the West—the gamblers, whores, thieves, and similar generic types.[43] Many of the first and second groups were men of the highest type who possessed both education and considerable business experience. It has been suggested that the presence of these men provided the Black Hills with excellent material for the building of settled communities.[44]

In the early period of the rush, nearly all the newcomers went first to the diggings on French Creek. From there they moved down the creeks and up the ridges. Gold was discovered at Sheridan on Spring Creek as early as July 1875, and at Pactola on Rapid Creek shortly thereafter. By August miners were working north to Whitewood Creek, Bear Gulch, and the Nigger Hills district.[45] The mining was by placer methods, and its practical limits were French Creek on the south and Whitewood Creek on the north.[46] Gold in some amounts could be found in practically any stream within this area. Custer, the city built on the site of the first gold discovery, was the largest city in the Black Hills. Hill City, Palmer Gulch, Friday Gulch, Sunday Gulch, Newton, and Rockerville sprouted placer camps. [47] In all these camps some made quick fortunes, but at the same time many made nothing.

Placer mining was the only type of mining readily available to the lone man without capital or technical knowledge. A certain amount of experience could be helpful but was not absolutely essential. Placer gold originally lay hidden in ores, but eventually erosion broke it loose and deposited it in streams, which in turn left considerable amounts of it in the stream beds and in adjacent gravels. To obtain the gold, it was necessary to remove the gravel, place it in some sort of open-topped container—a pan, rocker, or sluice—and run water over it. The water would wash away the lighter gravel and leave the heavier gold. It is a process more simple in the telling than in the doing and required enormous amounts of intense labor with no guarantee of remuneration. This type of mining could become highly technical and require extensive capital in its higher form—hydraulic mining. In its simple form anyone could try it, and the hopeful entrepreneur was in business with a pick, shovel, and pan. Most of these men made day's wages if they were lucky, but the few who struck it rich provided the lure that kept the others coming.[48]

The solid wealth that could provide permanent employment and build stable communities awaited the discovery of the great lodes of gold ore from which the placer gold had originally come, and the miners diligently sought these lodes. By ones and twos, in large parties, organized and disorganized, they combed the Hills—ever seeking the source of the vast wealth they knew existed. The luckier ones went north, where some of them found the lodes.

There is some doubt as to the first discoverer of gold in Deadwood Gulch. Some state that it was a party of miners from Montana led by Ed Murphy, who visited the area in the autumn of 1875 and made a strike.[49] Others claim that A. S. Blanchard and party from Custer made the first discovery September 6, 1875.[50] The great majority, however, give credit to the party of Frank Bryant, John Pearson, Thomas Moon, Richard Lowe, James Peierman, Samuel Blodgett, and George Hansen. These men first explored the northern Hills in August 1875, and John Pearson prospected Deadwood Gulch.[51] The party was forced to leave on orders of General George Crook, but Bryant and another miner named William Lardner returned in November. Bryant staked his first claim on November 8, 1875. The next day Lardner, Pearson, and the others found gold in Deadwood Gulch. They made their winter camp at the mouth of Blacktail Creek and called it Gayville.[52] In

December 1875 these men organized the gulch under the name "Lost Mining District" and elected William Lardner recorder.[53]

The winter of 1875–76 was extremely severe, and heavy snow made travel through the Hills highly hazardous if not impossible. The word of the strikes in the north reached Custer, and the prospectors there fidgeted in helpless anticipation. The thaw did not come until March, but then there occurred the greatest stampede of people ever seen in the Hills.[54] The city of Custer, which had 1,400 buildings and 10,000 people, was depopulated in a few days. It was said that only fourteen people remained in the town.[55] By June 1, 1876, eager miners had taken up most of the desirable claims in Deadwood Gulch, and 1,000 men were panning gold or trading in the hastily built cities that were strung along the creeks.[56]

Deadwood, the premier city in the area and one of the more famous places in the West, was laid out on April 26, 1876. It was a raw place. The covering pines from the hillsides became cabins and whipsawed lumber. The scars of excavation showed clearly near the creeks. Main Street, which meandered along the bottom of the canyon, was obstructed by the sluices and diggings of the miners. These men lived first in tents or wickiups made of poles with boughs and dirt thrown over them. Later log cabins made their appearance, followed by board structures made of hand-whipsawed pine at $150 per thousand board feet.[57]

By the summer of 1876 Deadwood was a roaring gold camp of some 25,000 people. From one end of the long gulch to the other stretched a narrow jerry-built city full of every conceivable type of humanity.[58] Its people were bound only by their own rules; the United States could regard them only as trespassers. The miners met and organized their own government, established courts, and maintained whatever law and order they could.[59] A good many things occurred that would never have been tolerated in better established communities. There were countless dance halls with houses of prostitution on the second floor. There were gunfights, brawls, and wide-open gambling. Wild Bill Hickok died at the hand of Jack McCall in Saloon #10. Calamity Jane cut a wide swath through the town. Wyatt Earp and Bat Masterson were visitors, and Sam Bass and his gang allegedly held up a stage in the vicinity. Sunday was the big day. From every camp men poured into Deadwood to celebrate. Street preachers competed with gamblers for space in the streets, while free-lancing ladies-of-an-evening elbowed them both out of the way in the race for customers. Chinese moved silently through the crowds dispensing and collecting laundry. The mail, if any, was delivered. The universal currency was gold dust. Oxteams and herds of cattle, trail-driven from as far as Texas, moved through the streets.[60] The atmosphere was lusty and exciting.

However, while Deadwood roared, the majority of the miners were working hard for short wages; indeed, one of the gulches was named "Two-Bit" because that was the amount a man could make for a day's work there. Although the first lode claims had been located as early as December 1875,[61] most of the mining was still of the sluice or placer variety.[62] An estimated $1 million was taken out of the Hills from this type of mining in 1876.[63] A relative few took it, however, for the average Deadwod Gulch placer claim paid little more than living expenses. A good many men in the camps were without money.[64] As a disgruntled visitor to the Hills observed, the gulch placer mining was about used up, and the quartz lodes were the only real hope.[65] He also felt that the gold rush, which had already boomed river traffic on the Missouri and triggered the Great Dakota Boom, would soon prove a failure.

This would be no failure. The prospectors working their way up the gulches had already found the greatest of the lodes, and other men had already started the meager little settlement that was shortly to house the miners who would dig out the great fortune. The lode was the Homestake, the settlement the future city of Lead. The richest of all the mines in the Western Hemisphere, perhaps in the world, was on the brink of development.

CHAPTER TWO

HOMESTAKE

THE PLACER MINER was capable enough to wash gravel but lacked the knowledge and skill to discover gold still encased in the rocks. He could look directly at a lode claim and be unaware of it. Thus lode men, moving up the gulches from Deadwood and probing the ground from stream to ridge, sought the key to deep-rock mining. They had been in earlier rushes, had seen lodes, and had perhaps worked hard-rock mines.

The men who found the Homestake lode were this type—practiced and skilled at their trade. The most successful were two brothers, Fred and Moses Manuel. French Canadians from Minnesota, they had knocked around the mining frontier since the end of the Civil War and had prospected and mined in Montana, Utah, Idaho, Arizona, and Nevada. Moses went to Alaska in 1874, thence to British Columbia, and finally found gold near Great Slave Lake. He was in Portland, Oregon, awaiting passage to Africa when he read of Custer's discovery of gold in the Black Hills and immediately wrote Fred, who lived in Last Chance Gulch (now Helena), Montana, suggesting that the two of them go to Dakota to try their luck.[1]

The Manuels left Montana and went to Custer City in 1875. They found nothing in the southern Hills and started northward, traveling through Hill City, across Box Elder and Elk Creeks, and down Spruce Gulch to the diggings on Whitewood Creek.[2] There they met Hank Harney and Alex Engh, who accompanied them to Deadwood Gulch. Here each of the four staked a claim at the mouth of Bob Tail Gulch.[3] These claims were eventually consolidated in the successful Golden Terra Mine, but the men were not satisfied and continued to search. They eventually worked their way up Gold Run where they helped J. B. Pearson locate the Old Abe Mine.[4]

They wintered near the Old Abe and in the spring of 1876 made their big strike. Moses Manuel, over the objections of his brother, was looking for gold while the ground was still snow covered. The snow was melting, and the water ran down a small draw. At the bottom of the draw Moses saw some exposed quartz.[5] He turned to Harney and said, "Hank, this is surely a homestake."[6] ("Homestake" meant sufficient capital to establish one permanently in the states.) Moses Manuel describes what then happened:

> Next day, April 9, 1876, Hank Harney consented to come and locate what we called the Homestake Mine. We started to dig a discovery shaft on the side of this little draw, and the first chunk of quartz weighed about 200 pounds and was the richest ever taken out. We came over the next day and ran an open cut and saved the best quartz by itself. Afterwards we built an arastra[7] and hauled the ore over. We ran the arastra the following winter and took out $5000. That spring we sold the Terra to John Baily of Denver and Durbin Brothers of Cheyenne for $35,000.
>
> The next work we did was to bond the Homestake to a California company for $40,000 and the Old Abe to Wostom Bros. for $5000. Brother Fred and Hank Harney took a trip to Chicago and I stayed to work the Homestake alone. I put up a ten stamp mill and bought a half interest in Gwinn's stamp mill. I ran the ore from the Homestake through this mill; all the other mills in the neighborhood were running the Homestake ore at this time. When I had a spare man more than I could work on the Homestake I would put him prospecting the Old Abe chute of ore. By that time the two bonds had expired and no sale had been made. Fred and Hank Harney came back and we had improved the property so much we concluded it was worth more than the bond and we wouldn't sell for any such figure.
>
> L. D. Kellogg, the agent of Senator George Hearst, came up one day and wanted to get a bond on the Homestake and we agreed on a bond at $70,000 for thirty days. A few days later Captain Huron came up and wanted to buy Old Abe, and offered up $45,000 for it which we took. Both bonds were complied with and paid for within the limited time.[8]

The Manuel brothers had discovered and sold the principal gold producer of the Black Hills and perhaps the world—the great vein that was to prove as nearly inexhaustible as is conceivable.[9] It was formed in an immense roll or fold of the schists and dipped to the southeast and from the horizon at an angle of about thirty-five degrees. It averaged about 430 feet in width.[10] At the time of its opening the ore in the mine was worth $14 per ton gross.[11] This was not abnormally rich ore; in fact it was rather low grade, but the quantity was huge. The ore itself was a simple type that was admirably adapted to stamp milling and amalgamation on a large scale.[12]

The Manuels had found their homestake and achieved what to them was considerable wealth, certainly far more than most prospectors achieved. The great wealth, however, would go to others. This was not a case of the wicked capitalists coming in to rook the poor-but-honest prospector. The Manuels were satisfied with what they received and rightly so. They were selling a claim that covered 75 feet on each side of a little draw and extended 1,350 feet from the middle of Gold Run to the top of the ridge.[13] On this claim they had driven a small discovery shaft and done some open-cut work, but the property was not in any sense a developed mine. To make the Homestake pay would require capital on a large scale, and capital invested in unproven gold mines is capital exposed to the greatest possible risk. In addition to the financial investment, it was necessary to apply a high degree of technical skill and organizational ability to this mine. Unusually large quantities of this ore would have to be treated by machinery and chemicals in the mass. Such technical and organizational skill was as unobtainable to the prospector as was large-scale capital. The Homestake, to be operated effectively, had to be run as big business.

The Homestake was purchased not exclusively by Hearst as Manuel stated but by a syndicate headed by Hearst. The other members were Lloyd Tevis, a partner in the Central Pacific Railroad and president of Wells-Fargo Express Company, and James Ben Ali Haggin, also in Wells-Fargo and other Pacific Coast business ventures.[14] Hearst and Haggin were universally considered to be plungers, but Tevis was known as a man of sound judgment. When he invested, it was evident that considered thought and investigation had preceded the venture.[15] The three were a remarkable combination. Hearst had made his first sizable money in the famed Ophir Mine in the Comstock Lode. In 1872 the trio purchased the Ontario, the richest silver mine in Utah. In 1881 they would purchase the Anaconda in Montana at the behest of Marcus Daly. This would be the greatest of all American copper mines.[16] They apparently reduced mine purchasing to a science.

The Hearst syndicate was most interested in the news of the Black Hills gold rush, and this interest rose with reports of the advent of quartz mining. In June 1877 they dispatched L. D. Kellogg, a practical miner of the type preferred by Hearst, to the prospected region in the vicinity of Lead. He was clothed with discretionary power to negotiate the purchase of satisfactory properties.[17] Kellogg had an early chance to buy the Homestake for $50,000 cash with no "bedrock"—no deferred payments to be made from the earnings of the mine —but turned it down.[18] The Manuels were exceedingly short of cash at this time. They made some money from the mine of course; and Harney sold cordwood to other miners; but the expenses for steel, explosives, food, and help ate this up quickly. The brothers forced 100 feet of their claim (one-fifteenth) on H. B. Young, a Deadwood merchant, in exchange for food and supplies. Later, Kellogg was quite content, if not happy, to pay $60,000 for fourteen-fifteenths of the mine.[19] After enjoying a short partnership with the San Francisco nabobs Hearst, Tevis, and Haggin, Young sold his interest to them for $10,000.

Upon bonding the Homestake, Kellogg went posthaste to Sidney, Nebraska, where he caught the Union Pacific for San Francisco. There he reported the purchase to Hearst, who returned at once with him to the Hills where the final acts of the transaction were completed.[20] Before returning to San Francisco, Hearst also purchased the adjacent Golden Star mine. Thus the process of accumulating the whole of the great lode began early and would continue long past the turn of the century.

After his return to San Francisco, Hearst, together with his partners Haggin and Tevis, purchased an eighty-stamp mill and shipped it to Sidney by Union Pacific. From Sidney it was transported to Lead City, the home of the Homestake, by bull team. The bulls perished in the snow, but the mill was dug out and taken to the mine site—no doubt by stronger and warmer bulls. There it was erected and ready for operation by July 1, 1878.[21]

Another task undertaken by George Hearst was the establishment of the corporate structure for the new mine. The loose partnership arrangement employed in the purchase of mining properties would not serve for long-haul development. As a consequence, the Homestake Mining Corporation was incorporated under the laws of the State of California on November 5, 1877.[22] The major assets of the company were two claims, each 1,350 feet long and totaling 450 feet in width. The large amount of profitable ore available made it imperative that another mill be added, and one of 120 stamps was begun on December 1, 1878, and completed September 1, 1879. By that date the company had $380,833.78 invested in stamp mills. The debt for the mills was extinguished out of monthly receipts.[23] The company was profitable from the beginning. The first dividends were paid on January 15, 1879, even though the corporation purchased the adjacent Nellie claim. By April 1, 1880, the mine had produced $1,305, 903.39 in bullion and paid $450,000 in dividends,[24] despite the fact that the mills had to be stopped in the winter of 1879–80 due to extreme cold.

The exact ownership of the company in this early period (and later for that matter) is difficult to ascertain. Apparently Hearst owned about half the stock, Haggin most of the rest, and Tevis a relatively minor share.[25] There never seemed to be any doubt that Hearst really ran the company, although he never held office in it. Haggin was the president until his death and took an active interest in the Homestake, but Hearst called the shots. The Homestake gold was the foundation of the Hearst fortune, built on mining and enlarged by the newspapers of his son, William Randolph. Hearst had money before he bought the Homestake, but it was running thin; he had even had to sell his wife's carriage horses. But he would never be short again. The gold would buy into Anaconda, Mexican silver mines, the great San Simeon cattle ranch, and a seat in the United States Senate. The original investment paid off handsomely.

The Hearst method of running the Homestake was both simple and direct. The Senator merely hired the best man available, preferably one schooled by experience and little else, and gave him almost total autonomy in superintending the operation. The only real condition attached was that a large profit be made; with an ore body such as the Homestake possessed, this could be complied with admirably. This is not to deprecate the company's superintendents. They were an outstanding group of men and, with one exception, invariably successful in overcoming difficulties. The first of these superintendents was an Ulster Irishman named Samuel McMaster.[26] He had been mining for ten years and was a veteran of the California and Australian gold fields. Hearst sent him from San Francisco to Lead City in 1877.[27] When he arrived, the Homestake was two narrow claims with a few shafts started in them. He had to place the equipment and start the mining operations.[28] He supervised the acquisition and building of the vast water system, acquired timber resources, organized the railroads, sank shafts, ran drifts, and supervised the recruitment of the first labor force. He was a mass of driving energy who succeeded in developing the permanent mining industry of the Black Hills.[29] His fantastic forcefulness made him a man admired, not loved. George Stokes, who was associated with the rival Father DeSmet Mine, accused him of diamond drilling other claims on the sly to determine their value and of sloppy mining technique that led to increased production but also to slides and cave-ins.[30] The truth of both charges is impossible to ascertain, although the areas of Lead under which McMaster mined did sink some 27 feet in the 1920s. McMaster did not stay long with the Homestake. He made an abortive attempt to run for the governorship of Dakota Territory and died shortly thereafter in 1884.[31]

McMaster's successor was not a miner in a real sense; he was a former secretary and chief bookkeeper for the first superintendent.[32] Thomas J. Grier was essentially an administrator and organizer. He handled men well and knew how to choose competent subordinates and give them authority, a trait McMaster lacked. He took over a mine that was in full operation with the first vital development work completed. What was needed was competent administration, capable expansion, and technical advance. Grier was the man to provide these. His administration was far from tranquil and saw great troubles with labor and fights with rival companies over expansion. Grier did not confine himself exclusively to mining activities. He ran the company railroad (the Black Hills and Fort Pierre), supervised the company store (the Hearst Mercantile), and became president of the First National Bank of Lead in 1890.[33] He was extremely community-minded. In this he was amply supported by Phoebe Apperson Hearst, who had succeeded to the majority ownership of the company in 1891 upon the death of her husband.³⁴

Grier died rather suddenly in 1914. His successor was former assistant superintendent Richard Blackstone, an old company hand. Blackstone had been a captain in the Civil War and had studied engineering at Rensselaer Polytechnic Institute. Hearst had hired him to be chief engineer for the Homestake. In this capacity he had handled not only the mining work but also the railroad construction and the erection of the hydroelectric works in Spearfish Canyon.³⁵ Blackstone, who was a fine engineer and a competent assistant to Grier, was something less than a success as superintendent. The mine was trying to produce as much as possible at government request due to the war. Against expert advice Blackstone sank a new shaft—the Star—very close to the open cut. So close was it that it collapsed into the cut, much to the consternation of company officials, who subsequently removed Blackstone from office. As the only superintendent to suffer this indignity, he spent his remaining years in considerable bitterness.³⁶

The fourth superintendent of the Homestake was Bruce Clinton Yates, a man of charm and culture as well as professional ability. A graduate engineer, he had worked for the Burlington Railroad and engaged in private surveying before taking employment with the company as a mine surveyor. From that position he worked up to assistant engineer and then to assistant superintendent before taking over the chief job.³⁷ Yates wrote widely and well on engineering; he even dabbled in poetry.³⁸ Under Yates the company underwent extensive change in its mining techniques. The hallowed company practice of running anything and everything through its mills changed to a more selective method whereby higher-grade ore would be mined in a selective process, thus increasing the efficiency of the mills.³⁹ Yates introduced many safety measures, including wet drilling to lessen the incidence of silicosis. He also contended with labor shortages during the war and with decreasing profits in the 1920s. His sudden death in 1936 closed a remarkable career in the Homestake's service.

The final superintendent under consideration was Guy N. Bjorge, a mining geologist. Bjorge had experience in Arizona copper mining and had achieved great success in Latin

American mining as well as serving as a consulting engineer for many years. Like Yates, he spurred technical advances in mining and milling.[40] His greatest challenges would come in World War II, when the mine shut down at government order, and in the period of spiraling costs that followed the war.

The influence of the superintendents cannot be overstressed. They ran the company insofar as its mining ventures were concerned. They made the decisions on the spot and were almost invariably backed by the officers in San Francisco. The ultimate success or failure of the operations lay with them; the relationships with labor, the state, and the community were nearly exclusively theirs. This autonomy probably lessened somewhat through the years as the company settled into more established modes of business procedure.

The most important facts about the Homestake mine were that it invariably produced bullion and invariably made money. To a society that equates wealth with the ownership of gold mines, this may not seem unusual. However, the unvarnished fact is that gold mining is a business fraught with danger and normally possessing the steady growth potential of a roulette wheel. It is possible to get rich quickly, although the odds go the other way. The chance of being able to plan for a constantly growing company for a period of fifty years is practically nil. Veins pinch out, capitalization is either inadequate or overextended, labor makes operations unfeasible, inflation or deflation catches the miner in an exposed position, and the enterprise is over. The Homestake avoided all this and in the 1960s remained the only major gold producer in the United States.

The mine continued to be profitable. By 1880 $450,000 in dividends had been paid on a total production of $1,305,-903.39. This was done on the original claims with the highest-grade ore mined through open cuts.[41] Production remained fairly constant, as did dividends, until the 1890s. There were variations of course, but in general the production ran from approximately $1.1 million to $1.66 million, while dividends varied from $150,000 to $675,000.[42] The profits probably could have been higher if the company had not been engaging in extensive development work that required enormous capital investment.

By the 1890s the other mines in the area were experiencing

problems. The early phase of lode mining was passing as the high-grade ores near the surface were being worked out. The continuance of mining depended on sinking deeper shafts to work the plunging veins of lower-grade ore. This required capital investment on a large scale, and many mining companies lacked the resources of capital necessary. The Homestake had resources in abundance and started the final phases of buying out the smaller companies. From the beginning it occasionally had bought privately owned claims. In 1892–93, however, Homestake Mining Company spent $417,390 on new mining properties—many times the purchase price of its own mine and mill.[43] In 1893–94, the company spent another $103,780.[44] Through these purchases, the reserves of ore in sight went from an eight years' supply in 1893 to a twenty years' supply in 1895–96.[45] By such means the company acquired a near monopoly on all the ore in and around Lead, thus insuring the foreseeable future for itself and its employees. Still the owners kept buying. In 1899–1900 the Homestake spent $1,641,255.28 on property purchases—primarily the rather large Highland Mining Company with a 140-stamp mill, the Black Hills Canal Company, and the Black Hills and Fort Pierre Railroad.[46] The investment began to pay very quickly. By 1899–1900 the bullion production was up to $4,208,346.90 with dividends of $1,175,000.[47] By that time the capitalization had been increased from the original $10 million to $21 million and would be increased in 1902 to $21,840,000.[48]

The acquisition of the mining properties was not at all easy. Much of the success or failure of any of the mines depended on the ability to acquire the vast quantities of water necessary to the mining process. Obviously the company controlling the water had every chance to consolidate the mining industry in a given area. Water was a prize worth fighting for, and fight they did. In the early days the most serious rival to the Homestake was the Father DeSmet Mining Company, which was situated above and to the north of Lead near the back of the open cut at Terraville. An independent water company named the Foster Ditch ran its ditch by the DeSmet. The Foster Ditch was in serious financial difficulty and for sale to the highest bidder. The DeSmet was the logical purchaser but felt that its financial condition, due to expansion, did not warrant the purchase at that time.[49] At this point Sam

McMaster, with dubious motives and pliable ethics, stole the march on the DeSmet and purchased the water ditch from Foster. The purchase appeared to be of little or no value to the Homestake, which could not use the water without installing an elaborate and expensive pumping system.[50] Gus Bowie, the DeSmet superintendent, was not a man to be put upon. He retaliated by buying the Boulder Ditch, which ran close to the Homestake. It was much lower than the DeSmet works and thus of no value to that company except for bargaining purposes.[51] An attempt to compromise failed, and the resulting comedy was known locally as "The Great Water Fight."

The opening shot in this struggle was fired by the DeSmet. Bowie, afire with righteous indignation, informed McMaster that dirty water from the Homestake tailing flume was polluting the Boulder Ditch. McMaster hid his shame and chagrin well and countered by claiming that the DeSmet had no rights in the Boulder Ditch. Under U.S. mining law, anyone claiming right to water had to use the water to retain the right.[52] The DeSmet had not used the water because it could not use it. In desperation Bowie offered free water to the City of Deadwood in order to retain the water rights. Deadwood was delighted—free water is a joy to any harassed city treasury. There was, however, a complication which the Homestake promptly exploited. Another ditch company, the City Creek Water Company, had a franchise to furnish water to Deadwood. Aided and abetted by the Homestake, this company brought suit against the Boulder Ditch to enjoin it from giving its water to the city. In order to bring the truth and the light about the issues to the citizens of Deadwood, McMaster bought the *Deadwood Pioneer*.[53]

The civic authorities of Deadwood were caught in the middle of a power struggle and passed the responsibility of decision to the people. Both sides agreed that the citizens of Deadwood should vote on whether to accept the free water. The election was glorious; water was the issue, but alcohol was the means. Both sides furnished copious amounts of whiskey, wine, brandy, cordials, and beer in an attempt to woo the voters. Some teetotalers, immune to the charms of John Barleycorn, had to be bribed.[54] Some men with no ethical standards at all sold votes to both sides.[55] The Homestake with its usual efficiency knew when it was ahead in the balloting and succeeded in closing the polls early.[56]

Bowie, however, was not yet defeated. He acquired a neglected mining property on Split Tail Hill and put Pete McDonald to work washing gold there with water from the Boulder Ditch. No one ever saw any of the gold produced there, but they were using water. At the same time, the DeSmet established a newspaper, the *Evening Press*, and with George W. Stokes as editor sallied forth to do battle again.[57] Stokes also found another use for the water; he sold some to a laundry. The DeSmet again brought suit against the Homestake for polluting the water. McMaster built three settling dams to clear it. By this time every lawyer in Lead and Deadwood was working on the court case.[58] It would seem that pollution of the water did occur, but the courts were unwilling to close down the Homestake over the very minor Split Tail workings. Judge Gideon C. Moody, for years a noted Dakota Territory politico and a member of Yankton's "Broadway Gang," rendered the decision; it came to be recognized as the best standard on riparian law as it applied to the Black Hills.[59] Judge Moody later resigned his judicial post and became chief legal counsel for the Homestake. With the company's backing he became prominent in territorial politics and joined his employer, George Hearst, in the United States Senate when South Dakota achieved statehood.[60] This should not imply any conflict of interest on the water rights case, however, for the law was sound and repeatedly upheld.

The final settlement of the water case required that the Homestake return the ditch to Foster while the DeSmet would turn over Boulder Ditch to the Homestake in return for $30,000.[61] There can be little doubt that the outcome favored the Homestake. Control of the Boulder Ditch gave the company a stranglehold on the water from Whitewood Creek, the major supply in the area. The struggle also spurred the company toward the acquisition of nearly all water rights in the northern Black Hills, which rights would be put to use for purposes of water itself and hydroelectric power. This was a key factor in the dominance of the Homestake.

The expansion continued after the turn of the century as the company bought out its competition on the ore body. In 1902 the Black Hills and Fort Pierre Railroad was sold to the

Chicago, Burlington, and Quincy for $1,091,037.40.[62] In the same year the Deadwood-Terra mining complex, which included the DeSmet, came under company control, as did the Caledonia mine.[63] This completed the consolidation of the big mines on the Homestake belt. The company held some 350 claims and a total area of 2,624 acres.[64] In 1912 the company added to its holdings in the Whitewood mining district by purchasing the Columbus and Hidden Fortune mines; in 1925 the Oro Hondo property was added. By the mid-1920s the aggregate mining holdings had gone to 577 patented mining claims for a total of 3,563 acres.[65] This does not include the coal lands (mainly in Wyoming) and the timberland, held either in fee simple or on a lease basis.

The ultimate purpose of the purchases was to give the Homestake control of the entire ore mass in the region. Centralized operations on the single source of raw material were necessary if that source were to be used to its utmost efficiency. Sooner or later, one-company control was probable; if it were not the Homestake, it would be one of the others. Such control was not only advantageous from the standpoint of profits but without it, the longevity of the mining industry—nearly the sole source of wealth to the entire region until the advent of tourism in the 1920s and 1930s—would be seriously impaired to the detriment of all. Unilateral management gave sufficient efficiency and continuity to fully exploit the mineral potential of the region. The Homestake won its control through sound leadership, strong financing, luck, and ruthless determination.

The combination of the new properties and recent technological advances, primarily the cyanide process, increased the revenues of the company. In 1902–3 the company produced $5,055,506.32 in gold bullion with dividends of $819,000. The efficiency becomes apparent when it is realized that this was done with ore that averaged $3.539 per ton.[66] The key was production in mass. By 1903–4 the company had 1,000 stamps pounding in the various mills and two of the largest cyanide plants in the world.[67] Homestake was mining on the 1,550-foot level by 1905–6 and constructing a slime plant in Deadwood to squeeze the last bit of gold from the ore that was as fine as face powder by this stage.[68] According to the usual practice by this time, twenty years' ore was blocked out in advance.

By 1914 the company was producing $7,736,970.17 per

annum in gold and paying dividends of $2,210,208. That year saw major changes in management as both Superintendent Grier and President Haggin died. Haggin was succeeded by Edward H. Clark, a member of the Board since 1894.[69] Clark, who represented his relatives, the Hearsts, would remain in the presidency until 1944 and run the San Francisco offices and the Board with an iron hand.[70] By 1917 the company at government request was producing tungsten at cost for the war effort and attempting to increase the gold produced in the face of a rather severe labor shortage that continued into the 1920s.[71]

In 1923 the greatest threat to the life of the mine came to light. It appeared that the great vein was pinching out, and the supply of ore, hitherto considered inexhaustible, would no longer be available. There were 16,357,168 tons of ore blocked out and 901,549 tons broken down and awaiting processing in the mills.[72] By 1924 the reserve lessened to 15,815,118 tons blocked out. It appeared definite that the days of the mine were numbered. Officials of the company debated on what should be done. One theory suggested that the only sensible solution would be to mine and mill the available ore and close down the operation when that ore was exhausted. The other, expounded by Donald H. McLaughlin (who had done his Ph.D. dissertation at Harvard on the geology of the Homestake),[73] considered that the veins had not pinched out but rather had suddenly plunged. As mining is done on strict horizontal and vertical planes, this was not apparent to the miners. The only way to find out was by extensive diamond drilling at enormous expense. If the veins were pinched out, this expense would be for naught. It is on such issues that executives and boards are put to their ultimate test. In this case they guessed correctly and ordered the drilling which proved McLaughlin's theory to be correct.[74] By 1931 ten years' ore was blocked out, the development on the lower levels was encouraging, and dividends of $2,122,307 were paid. The ore's value had also risen due to Yates's selective techniques of mining; the company was now able to produce ore worth $6.36446 per ton.[75]

The great depression brought economic disaster to most Americans, but it helped the Homestake and its laboring force. The New Deal was anathema to the company politically, but economically it proved a godsend when it devalued the dollar. This devaluation took the price of gold from its usual $20 per ounce to $35 per ounce in 1935. It also brought some side effects that would hurt badly in the 1950s and 1960s (an absolute ceiling on the price and the limiting of sales to the U.S. government), but these were not apparent in the 1930s. In 1935 the total value of gold production had risen to $19,661,-642.56 while operating expense stood at $7,929,274.97—a net profit of $11,732,367.59. Even after paying a federal income tax of $2,156,503.84 and reserving $3,587,839.27 for depreciation, the company profited enormously.[76] Wages had gone up, but expenses for supplies such as steel, powder, and machinery went down. Lead was an oasis of plenty in a desert of misery. The general prosperity continued until War Production Order #L-208 closed the mine on October 8, 1942. Franklin Delano Roosevelt had different views on the desirability of gold production than had Wilson in an earlier war.

The years from 1877 to 1942 had seen great profits from the single mine run as a monolithic enterprise by a single company. During this period the Homestake had continually solved the major problems facing it and had achieved stability and longevity on a scale unprecedented in American gold mining. The mine had shut down on only three occasions in all that period. There had been a mine fire in 1907, labor troubles in 1909–10, and another fire in 1919–20,[77] when the working force was kept on the payroll and worked on the surface.

The major problems besetting any producer of gold are essentially those of production and mining costs. The producer never has to seek a market or exert any effort to sell his product. He has no control over the price which is decided upon by either the world market or his government, but it is usually a stable price and can be relied on. In order to make a profit, he must insure that his costs of producing an ounce of gold stay below the price of that ounce of gold. Since he has little control over the price of dynamite, machinery, and the various supplies necessary to the mining and milling process, the only variable available to him is the cost of labor. This is not very much of a problem in a mine of high-grade ore, but in a mine of low-grade ore it requires highly skilled management and technology to achieve a profit gap. When that gap closes, labor is inevitably squeezed, and severe labor trouble results. The Homestake achieved, maintained, and expanded a profit level that not only aided its stockholders but also guaranteed that labor could be treated in a fair manner. This financial stability plus the long-term employment prospects would be the major factors in recruiting and maintaining a stable labor force and in deciding policies in regard to labor.

CHAPTER THREE

THE WORKERS

THE GREATEST GOLD MINE in the world has no value unless there are men to work it. These men, in the pit and in the mills, can have the most serious effect conceivable on the success or failure of a mining venture. The miner in particular can cause either, for he is no ordinary industrial worker. He is not one who can be set a simple task to endlessly repeat. Whether skilled or not, experienced or not, competent or not, he must exercise individual judgment. He does not work in gangs but in small teams wherein his is an integral part that is hard and frequently dangerous. He is of course supervised and subject to regulations; yet in the final analysis, he is worthless if merely an industrial robot. The nature of his work makes him an individual who must place the utmost faith and reliance on his fellow miners.[1] He trusts his own kind and is somewhat suspicious of those who work in the blinding glare of day on the surface—a suspicion compounded with envy, for the miner has a love-hate relationship with the drifts, raises, and stopes below the surface.

Labor relations in the mining industry have usually been precarious at their best and downright bloody at their worst. When the miners clash with the owners—be they coal miners in a British colliery[2] or copper miners in Montana—the frequent result is violence. Nowhere is this more true than in the precious metal mines of the American West. It is not clear why this should be so, but the evidence of the occurrence of violence seems incontrovertible.[3] Speculations concerning the effect of the hard, dangerous work probably have some merit; yet the question of the cause of the miner's violence—as is the case of most questions concerning the ultimate motivation of man—is perhaps unanswerable.

The men who worked for the Homestake came from

29

many lands and represented many cultures. In the early period of the mine, under the Manuel brothers and during the early superintendency of McMaster, the workers were men who had failed to strike it rich in the rush and were forced to work for wages in order to get a grubstake, to move out, or simply to survive.[4] The man who went on a gold rush was an entrepreneur, a man on the make, and failure gave him few alternatives. Assuming he stayed as a worker and did not make a strike with a grubstake painstakingly earned by the sweat of his brow, he found his status drastically changed. From the high plateau of the incipient millionaire (as he must have seen himself), he had to face the reality of a lifetime of working in the quartz mines. It was quite a comedown, and many had difficulty adjusting. Even if the adjustment came easily, the prospector would still be faced with the problem of entering a difficult trade with no particular preparation.

The establishment of the Black Hills mining frontier as a solid quartz mining region with companies meeting payrolls meant, among other things, that the professional miner would move in. He did not rush from hither to yon and back again seeking immediate riches but rather worked the hardrock mines for a living—a man skilled in the peculiar tasks of the miner. Of these, the most skilled and most experienced were the men of Cornwall—the "Cousin Jacks." Their appearance in number at any mining camp meant that the fly-by-night period was over and the camp gave strong indications of being a steady producer.[5]

The Cousin Jacks were the end product of centuries in the mines. In their native land, the southwest corner of Britain that culminates in Land's End, they had operated the tin mines since the times of the Phoenicians. By the eighteenth century Cornish mines extended out under the sea itself; by the middle of the nineteenth century, however, the mining was becoming inefficient. The mines were old and worked out.[6] Thus the Cornishmen migrated in ever increasing numbers to areas of the world where new mineral strikes were being made. They could be found in Australia, South America, Africa, and the United States. They came to Lead when they heard about the Homestake, its good wages, and its working conditions.[7] Many came directly from England; but many also came from other mining regions in the United States, particularly the copper

mining regions of Michigan's upper peninsula, where the pay
was bad and the mines were wet and dangerous.[8] They brought
with them the priceless techniques of the miner's craft. Anyone felt lucky to be teamed with a Cousin Jack, because he
knew the most efficient way to perform the work and his presence was an enormous safety factor. The Cornishmen could
read the rock and tell if it would be likely to collapse; their
keen ears were carefully attuned to the creaking of ore from
which they could detect signs of danger. Quick and intelligent,
they frequently assumed the leadership of their fellows.

The Cornish were not the sole members of the early working force. There were others from Great Britain—Scots, Welsh,
and English—and a large number of Irish. The most numerous
were those born in America.[9] Nearly all had roots in the "old
immigration" from northern and western Europe and tended
to be Protestant, with the exception of the Irish. Nearly all
spoke English which made assimilation into the work and the
community easy.

As the "new immigration" began to take effect in the mid-
1880s, the composition of the Homestake employees began to
change somewhat. New racial and ethnic groups started to
come to Lead to find work. Among them was a sizable number
of Italians who came primarily from the Piedmont of northern
Italy.[10] They were not miners by profession and normally
started in the mines as common laborers. They were faced
with enormous barriers in terms of both language and skills;
it would be years before either were acquired. Assimilation
came slowly as the Italians clung to their European customs
and settled together in the section of Lead known as Sunny-
hill.[11] With other Catholic groups they formed a solid basis
for the church in Lead.

Possibly the most colorful of the varied nationalities were
the Slavonians. Primarily Serbs and Croats, with an occasional
Montenegran, these people had been citizens of Franz Josef's
polyglot empire. Today they call themselves Yugoslavs. They
were a people divided by language and religion. The Roman
Catholics, mainly from the Dalmatian coastal area, spoke Croa-
tian; the Greek Orthodox peoples from the interior spoke
Serbian and usually became Episcopalians in Lead, as did the
few Moslems among them.[12] For the most part they came directly to the Hills from the old country. Like the Italians,

they preferred to live together along a narrow twisting street called Gwinn Avenue, known locally as Slavonian Alley.[13] They tended to assimilate very slowly and lived unto themselves. They imported marriageable girls from their native areas through marriage brokers and maintained their foreign customs longer than most other groups.[14] So anxious were many of them to have employment at the Homestake that they would come and await their chances for as long as a year, living in the meantime on credit from the innumerable Slavonian boardinghouses in the alley.[15] When finally employed, they proved to be strong, willing workers.

There was a large Scandinavian community in Lead working for the Homestake. This was not unusual in Dakota, which underwent extensive Scandinavian immigration in the 1870s and 1880s.[16] By the late 1890s the combined population of Swedes, Norwegians, and Finns in Lead was probably as large as that of the Cornish.[17] The most numerous of the three were the Finns, who had migrated from Finland first to the copper regions of Michigan or the iron mines of the Mesabi Range in Minnesota and thence to Lead. They maintained their group cohesion and settled together high on a mountaintop above the Yates shaft. They retained some of their old customs, such as their penchant for steam baths—the sauna. They also retained a reputation for political radicalism that would affect their relations with the company, the union, and the rest of the community. The Swedes and Norwegians blended into the community better than the Finns. They had no distinct pattern of settlement or any of the distinctive national societies enjoyed by the Italians, Cornish, Slavonians, Irish, Scots, and Finns.

The other racial and ethnic groups in Lead lacked the group spirit of those previously mentioned. These included Germans, Russians, Canadians, Spaniards, Swiss, Turks, and Dutch. The non-Europeans—Negroes and Chinese—worked mainly as menials and small entrepreneurs. The Chinese ran laundries and restaurants. Neither had any representatives working in the mines.[18]

To a certain degree, all the racial-ethnic groups working in the Homestake mine were undergoing assimilation into American culture and society. The process was seldom totally effective and varied widely from group to group. Thus a

Cousin Jack, whose language problem was nil and who was quite used to a similar culture and comparable political institutions, assimilated very easily. A Finn, on the other hand, had a severe language problem and a totally different philosophy of society and government. Assimilation would take a long time for the Finns, Slavonians, and Italians. Even among the most easily assimilated, certain racial-ethnic considerations remained and were liable to come to the fore at any time. The Irish could and did adjust quickly and yet remained Irish. The proper stimulus, perhaps something connected with the British Empire, caused a reaction that was purely Irish. In Lead the mines, mills, and the union would mix the nationalities together; all men would be roughly equal and roughly similar in status. After work, however, each would go home to an area composed of his own countrymen, buy his groceries at a store run by one of them, visit a saloon that catered to them alone, speak his own language, and play the old games. Most were hyperpatriotic Americans; yet they retained an identification with the area, the culture, the society they had voluntarily left. They could exhibit fierce pride and fierce shame at the same time about a background whose influence diminished slowly but seldom completely. Racial and ethnic considerations were always evident in Lead; in the 1890s two-thirds of the Homestake employees did not even speak English.[19]

The Homestake men were under various types of supervision and direction in the mines and mills. The management in the mills, machine shops, foundry, and refinery was similar to that of any other industry. Each unit had a department head and under him a foreman who conducted most of the operations. In the mine, however, there was a different system, although a superficial similarity existed. After the first few years when the superintendent personally served as head of everything, an official was in general charge of the mining operation itself. He had under his immediate control the professional engineers and technical personnel as well as the skilled labor. The miners were actually supervised through highly empowered shift bosses, who always retained great authority over the men, although this authority diminished as time passed. Until 1904 when the company established an employment department, the shift bosses had the power to both hire and fire the men on their crews.[20] The hiring process was in some ways similar to the shape-up employed by longshoremen, except that the shift boss would usually have a steady crew that worked for him. When a man was fired or quit, the boss (if he had not made a previous arrangement) would go to the lines of men seeking employment and choose one.[21] The power to fire was just as arbitrary. The shift boss could discharge any member of his crew for no stated reason or could exercise his option to use a milder form of discipline—usually an enforced leave for a stated number of days. This unwanted vacation was without pay and would remain the basic form of discipline until the present.[22]

The individual shift boss was not necessarily unreasonable in his judgments, and there were reasons for giving him considerable power. He had charge of mining on a definite level or section of the mine. This implied that he was responsible for the basic elements of production for the entire company, and the production standards had to be met in order to make a large-volume mining and milling process pay with low-grade ore. In addition he was responsible for the safety of his men; the most frequent cause of "giving days" was a breach of safety rules.[23] The failure to "bar down" the ceiling of a blasted stope properly could cause its collapse and death to those caught beneath it. Careless handling of explosives, railroad iron, tools, and machines could also mean maiming or death.

The shift boss had many inherent problems in his handling of men. The individualism of the miner and his steadfast belief that he was right in his professional judgments meant that a strong and capable hand was necessary to control him. The language problem was gigantic. With 75 percent of the working force unable to speak English, it must have been a veritable babble underground as the orders were transmitted to the men with varying degrees of accuracy. Add to this the constant flow of spot decisions the shift boss had to make without help or advice, and it is easier to appreciate the problems of the job and the importance of the man.

The shift boss seldom had any formal training for his job. Then and now he was recruited from the ranks of the working miners and chosen for qualities of professional excellence, character, and ability to both handle and earn the respect of the men under him.[24] In the years preceding the first World War, he would almost certainly be a native-born American, a Cousin Jack, an Irishman, or a Scot. These choices were not due to any particular racial prejudice on the part of the higher authorities, who were mainly concerned with finding the best man. Immigrants from other areas were usually disqualified because of language, lack of experience, or inability to handle men. When these handicaps were overcome, there would be Finnish, Slavonian, and Italian shift bosses.

The subjective processes of selection of shift bosses worked well enough for the most part. Yet the power given with the position undoubtedly corrupted some, and there were no doubt rank abuses of authority. It is likely that some shift bosses took a rebate of their men's pay in consideration for hiring them in the first place.[25] This was the major reason for establishing the employment office in 1904. It is also probably true that authority was somewhat abused underground. The only recourse the individual worker had was that of appeal to higher authority up to and including the superintendent. This was encouraged by the higher authorities,[26] but it took a brave man to try this, and few did. Regardless of the sincerity of the higher echelons in promoting an open-door policy toward the ordinary worker, the communication problem would always remain. It would not always be a source of trouble and seldom affected large masses of men but occasionally resulted in individual injustice.

Foremen of the surface facilities of the Homestake shared in like measure the powers of the shift boss. They too could hire and fire until 1904 and were always able to exercise disciplinary action upon their men. Differences are apparent, however. For one thing, the immediate superior in a machine shop, mill, or refinery was much nearer at hand; thus disciplinary action came under his immediate surveillance more quickly. At the same time, the working force was smaller and composed largely of skilled and trained men who would be difficult to replace and who could more easily find employment elsewhere. The miner, on the other hand, would probably be classified as semiskilled labor; in a large underground operation like the Homestake, there were always plenty of men willing to take any discharged man's place. Unskilled and untrained laborers would be happy to move up to being muckers, who were better paid, and the latter were always eagerly awaiting promotion to miner. Thus the surface foremen did not wear as hard on their men as the shift bosses.

The Homestake working force consisted of 2,200 men in 1901.[27] The growth was fairly steady, getting its greatest boost from the opening of deeper levels of the mine and the extension of milling facilities. By far the greatest number of men employed were underground. The total number of employees would go as high as 2,800 but in general stayed about 2,500 until technology made it possible to produce increasing quantities of gold with fewer men. Even so, the employee level would never get below 2,000.

The most eminent authority on Western mining states that the Homestake pay scale was midway between that of the Comstock Lode and that of California, with skilled miners receiving $3.50 a day and surface labor and mill workers somewhat less.[28] Samuel McMaster reported paying miners $3.50 per day, laborers $3.00, carpenters and machinists $3.50, and mill hands $3.00–$4.00.[29] This would indicate that the millmen were paid at least as well as the underground workers. By 1905 the pay scale had changed somewhat. The unskilled laborers were receiving $2.50 per day on the surface and $3.50 below ground; the miners operating air drills received $6.00 per day. The latter seems little enough for handling the old-style piston drills that required both an operator and a helper at all times. The shift bosses received $12.00 per day,[30] twice the amount of the highest paid man in their crews.

Homestake mill labor was handled efficiently. The workers were paid more per shift than those in Colorado and California, but the relationship of labor costs to the amount of gold produced was much less due to the great crushing capacities of the mills.[31] In the stamp mills the foremen received $7.00–$8.00 per ten-hour shift. Head amalgamators got $4.00, amalgamators $3.50, crushermen $3.00, and laborers $2.50.[32] The surface workers may have received slightly less money than underground men by the turn of the century, but the difference was not appreciable or really commensurate with the amount of work, working conditions, or danger factors that the miner had to undergo.

The workers' standard of living was quite high for the period. They were paid by the day and worked seven days a week unless they requested Sunday off, which few did. W. P. Raddick, a pioneer and longtime resident of Lead, stated that an average family man could save money on this pay scale.[33] There is no reason to doubt this. The scale remained relatively unchanged until 1914 and was adequate to take care of the changes in the price index.

While good pay was important to the worker, stability and job security were just as important. The Homestake men had these to a greater degree than most hard-rock miners. The only layoffs of any consequence occurred in 1907 when the disastrous mine fire stopped work for a while, in 1909 when the labor troubles ensued, and in 1942 when the War Production Board ordered the closing of all gold mines.[34] There was a partial layoff due to water shortages in 1900; it affected 250 men for a very short time.[35] In 1902 the miners petitioned Superintendent Grier for a layoff in order that they might attend a ball. Grier granted this to most but kept the mills operating.[36] In general, the men worked as much as they wanted; more importantly, they knew they could work in the foreseeable future. The vast amounts of ore either stored in the mine or blocked out for future mining made it obvious that the company would be functioning for at least ten years in the future.[37] Few other mining concerns could offer this security, and it made employment at the Homestake highly desirable.

Prior to 1909–10 most of the workers at Homestake were union members. The Lead City Miners' Union was nearly as old as the Homestake itself. In the spring of 1877 the miners had combined "for mutual protection and for the purpose of securing for the men engaged in the hazardous occupation of mining for wages, a just compensation for their labors, and the right to use the fruits of their toil, without let or hindrance, or dictation from their employers, and to otherwise protect their mutual interest."[38] Pat O'Grady was the first president.[39] Approximately seventy men swore allegiance to the organization. The union, chartered in 1880 by Dakota Territory as a charitable and benevolent organization, was such in fact.[40] It paid up to $75 for funeral expenses of deceased brothers and provided color bearers and pallbearers for the funeral.[41] The union also paid compensation of $8 per week for a maximum of sixteen weeks to members who were unable to work because of illness.[42]

The union did very well financially and culturally. In 1878 it erected the first Miners' Union Hall. This structure, dedicated in 40°-below-zero weather, was outgrown by 1892 when the union passed a resolution to build a new hall at a cost not to exceed $25,000.⁴³ The new building was a splendid structure three stories high and constructed of variegated sandstone quarried near Lead. The first floor housed stores and shops on a rental basis; the second contained the largest opera house in South Dakota; the third contained the Hearst Free Library and meeting rooms for the union, the Ancient Order of United Workmen, and the Knights of Pythius.⁴⁴ The building was opened in December 1894 and was paid for by the original $25,000 and an issue of ten-year bonds. The 750 members were assessed $5 per month to help pay off the bonds.⁴⁵ Since the total cost was $70,000, the debt was formidable. The union, however, made extra money by renting the hall at $15–$90 per evening in a good capitalistic manner, and the burden proved not too great.⁴⁶ In addition to the rental properties, the union had a fine advertisement and recruiting device in the hall. It was the cultural center of the community. Touring thespians, sometimes of very high quality, played there. Nearly every large organization in town met there. All of this helped the union image in the community and with the workers who were not members.

The Lead City Miners' Union, benevolent and charitable though it might be, was dedicated to labor solidarity. In 1892 the union assessed each member $2 to help the embattled strikers at Coeur d'Alene, Idaho.⁴⁷ In 1894 it sent both money and tenders of any needed help to the Anaconda workers.⁴⁸ In 1895, at the request of the railway workers, the members contributed money to help pay Eugene Debs's legal expenses.⁴⁹

The position of the union in regard to the company was rather unsatisfactory from its point of view. There never was and never would be any kind of contract, written or verbal, between the two.⁵⁰ From the beginning the company was strongly committed to the principle of the open shop. This did not imply that the company at any time before 1909 interfered in any way with the union. The relations between the two were usually cordial and always polite. The company simply would not consider recognizing the union as the bargaining agent for all the men, although it would meet with

union officials. The Homestake did not object to its men joining the union, but it was willing to fight for the right of any
man not to join. The union made one attempt to get Superintendent Grier to approve of a closed shop.[51] He declined to
do this on the grounds that it would be unfair to the employees
who did not wish to join the union and because it would mean
a union voice in the management of the company. He was
friendly in his refusal, and the union made no particular dis-
turbance about its failure. It did continue to recruit Homestake workers and kept a list of scabs for some unspecified future action.[52] The closed-shop issue was the only real source of
friction; the union did not object to the wages or working conditions, and the question of hours was amicably settled when
it finally arose.

The major portent of possible labor trouble came in 1893
when the Lead City Miners' Union agreed to send five delegates to Butte, Montana, to meet with other miners' unions in
forming the Western Federation of Miners (W.F.M.).[53] The
importance of the Lead union in the W.F.M. is illustrated by
the listing of the L.C.M.U. as the second ranking local in the
W.F.M., exceeded only by the huge Butte Miners' Union. By
joining the W.F.M., the rather conservative Lead men allied
themselves with one of the most radical of all American labor
organizations—one whose leadership would espouse anarcho-
syndicalist doctrines and attempt to put them into practice.
The Lead union would provide much of the leadership of
the W.F.M. Charles Moyer, once a millman in Deadwood
and Lead, was the longtime president. James Kirwan, Richard
Bunny, Thomas Ryan, and J. C. McLemore were all prominent
in the W.F.M. This association was in strange contrast with
the peaceful unionism in Lead, where the only sign of the
W.F.M. was an increased drive for membership.

By 1899 an observer said that "the Miners' Union of Lead
had grown into a formidable combination of members, yet it
is gratifying to note that we never hear of 'strikes' and 'lock-
outs' or any kind of friction . . . in Lead."[54] By 1905 the Black
Hills Mining Men's Association, no disturbers of the social
order, spoke glowingly of the union's 2,000 members, its hall,
its lack of debt, and the surplus in its treasury: "It is one of
the models and one of the most conservative labor organizations in the world."[55] This was indeed high praise and would

be echoed in 1909 by *The Lantern*, the socialist paper in Deadwood, which said, "The Lead union has never been aggressive . . . it was a harmless charitable organization."[56] That the socialists and mine owners should agree on a union is unusual, and in a sense they were both right. The union was charitable, nonaggressive, and conservative. At the same time it was associated with an organization and a leadership that was the antithesis of these attributes.

The membership drive continued in sporadic bursts. In 1901 the union published the following:

> All non-union men and members in bad standing working in and around the Homestake mines, mills and shops are hereby requested to become members in good standing of the Lead City Miners' Union.
>
> These men are notified that a determined effort will be made by the Lead City Union, supported by the Deadwood, Terry Peak, Central and Galena unions, to have every man working in this camp a union man by 6th May. This movement has the endorsement of the Western Federation of Miners.[57]

The union was growing, although it did not achieve total membership by May 6. In 1902 Charles Moyer visited the Black Hills and found the union in a state of growth and prosperity.[58] Still, things remained peaceful at the Homestake. In March 1903 Grier wrote to the president of the union to complain that union officials had become somewhat overeager in their recruiting methods and had physically attacked a company employee. The union, on the same day, sent a committee to Grier to explain that they opposed violence, and the incident ended on the usual amicable note.[59] The union continued to tack up Join the Union signs and agitate for the closed shop in a mild way, but it was all done with goodwill, and general harmony prevailed.[60]

The union men achieved their only real victory in 1906, and it was not so much a victory as a gift. The unions of the Black Hills, as well as all over this country and even in England, had been agitating for the eight-hour day. The Lead union did not demand this for all members but only for those working underground. A union committee met with Grier to ask for this, and he stated that the step required study. A few days later Grier unilaterally issued the order for the eight-hour day for all Homestake workers, both underground and topside.[61] The union later claimed that Grier was forced to this by the threat of a strike.[62] The *Lead Daily Call,* writing at the time, said that Grier had been contemplating the step for some time and had only waited until the completion of the slime plant to put it into effect.[63] In any case, the eight-hour day came easily and was soon extended to the other mines in the vicinity.

By 1909 the working man at the Homestake enjoyed good wages, stable working conditions, good hours, and peaceful industrial relations. The previous difficulties between labor and management had been amicably settled. There were no outstanding differences on vital matters. Outward appearances indicated a calm and untroubled future.

CHAPTER FOUR

MINING AND MILLING

PROVIDING JOBS for workers and dividends for stockholders in a continuing and predictable manner is difficult in many industries but almost unheard of in American gold mining. In order to accomplish the feat, the Homestake had to develop and adopt processes essential to mining large bodies of low-grade ore profitably as well as organize the correlative activities necessary to support the primary work. Extensive coordination and high technical skills were required. In addition, a labor force of considerable size had to be trained and supervised. Although much of this activity was similar to that of any industrial complex, much was not. Mining is unique and requires techniques appropriate unto itself.

The mining of gold in lodes is complicated and requires several distinct steps. Lack of efficiency in any of these can jeopardize the entire system. Blasting ore from the surrounding rock formations is the first step. Next the miners move the ore to areas where it can be reduced to a size and state wherein the gold can be removed by mechanical and chemical techniques. Finally the liberated gold is refined to a pure state and poured into bars for final shipment and sale. The complexity of the whole procedure can be illustrated by a comparison with placer mining. In the latter, only the final step is required. All the rest has been done by nature. The gold has merely to be accumulated and refined—the simplest of all the steps. The lode miner, however, must discover the best techniques for every phase if mining is to remain feasible. It is undoubtedly true that in the American mining West, "the greatest achievement by far was technological—the art of mining."[1] Nowhere was this technology used to greater effect than at the Homestake, where the very life of the mine was dependent on it.[2]

The ore bodies of the Homestake lie in an extremely folded and altered bed of dolomitic limestone. The folding causes the ore vein to fluctuate widely in width from 50 to 400 feet.[3] The strike of the Homestake ledge is north 35° west and south 35° east.[4] The dip is irregular, and because the ore bodies pitch rapidly to the south, the northernmost mine on the lode (the Father DeSmet) first reached the bottom of the "pay shoot"[5] and ceased production in February 1921.

The gold-bearing ore is harder than the granite of Mount Rushmore,[6] and danger of collapse is relatively slight. The gold occurs in large bodies of quartzose chloritic schist, originally a ferrinous dolomite.[7] The gold particles are frequently small enough to be practically invisible.

The first approach to mining this ore body was by means of an open cut.[8] In the beginning the ore was removed from the top; later tunnels were dug on a level with the bottom of the gulch, and the ore was dropped through chutes[9] into cars in the tunnels. This method might have been used more extensively, but the miners—who lacked experience in working large ore bodies in such a way—were only too anxious to get underground to more normal mining.[10]

The managers of the Homestake mine were trained in the California-Nevada school of mining.[11] They rapidly put in the Comstock system of square-set stoping, a system devised by Philip Deidesheimer for the Hearst-owned Ophir mine. Within the stopes (excavations from which the ore had been extracted in steps) cubes were built of mortised and tenoned timbers; each cube was capable of being interlocked with others to the side, above, or below it. In the larger stopes some of the hollow cubes were filled with waste rock to form pillars for the support of the stope.[12] In stopes the mining always began at the bottom where a sill was cut across the ore body. Next the stopes were laid out and bulkheads of native pine constructed to support the pillars. The square sets were built as the sill floor was mined. When this was complete, the next floor was mined the same way. The mining proceeded upward until all the ore was taken out.[13] As the ore might continue for several hundred feet in height, the timbering was very impressive and exceedingly dangerous. The weight of the unmined ore became oppressive when the openings below were too large, and caving was the frequent result. This method required large amounts of timber; although the Homestake had its own sawmills and a train to bring in the timbers, the supply was not endless.

No system of regular stopes and pillars was devised until 1900, leaving much valuable ore unworked, never to be reclaimed. W. S. O'Brien, the general mine foreman, and Bruce Yates, then mine engineer, planned such a system.[14] Under this method the miners would start at the main vertical shaft and drive a horizontal tunnel, the main crosscut, across the course of the vein and through the ore ledge. Having intersected the vein, they would open a 24-foot drift from the main crosscut. This drift would follow the vein rather than intersecting it. Using the drift as a central passageway, the men then cut a 60-foot stope across the vein with 60-foot pillars separating the series of stopes wherein the ore was actually mined.[15] After the stopes were mined out, they were filled with waste rock or caved in.[16]

This original system was succeeded by the so-called Homestake system. As had been the case before, the tunnel was driven from the shaft across the vein. The miners removed the floor of the sill first and then put in timbered square sets with heavy planking, called lagging, over the top. They would proceed to mine upwards. As the ore was broken down from the roof, it would fall through spaces in the planking and fill the square sets beneath. When the sets were filled, each became a pillar that held up the stope. The lagging was then removed, and the floors above were mined without any further timbering. The advantage of the system was that the ore could be blasted loose without any fear of breaking down the timbering that held the stopes—they were held by large pillars of rock. The ore would be blockholed (drilled and blasted) on top of the piles and thus reduced to a size that could be handled. When the stopes were mined out, the crowns and pillars were mined in the same manner.[17]

The Homestake system gave way to the method called shrinkage stoping. The miners cut the levels and drifts in the usual way and laid out the stopes. Within the stopes they cut out the rock, by blockholing, from wall to wall and pillar to pillar. Next the men placed a floor of wooden posts topped with lagging. Two lines of the timber formed an area for trackways and for mucking the ore. The main problem was maintaining these floors; the timbering was frequently broken down by the weight of the mined ore. To prevent this, the company developed a technique wherein the miners undercut the pillars of solid ore on each side of the stope[18] and moved the trackways and the mucking operations to this area. These undercuts were 8 feet wide and needed no timbering because they were solid rock. Most of the mined ore was left in the stope until all the mining was finished. Because 6.5 cubic feet of solid ore occupied 10 cubic feet of space when broken, the excess ore was removed by hand-shoveling into ore cars.[19] The pillars were then mined with square sets. A 25-foot protective crown was left until last and then mined with square sets also.[20]

These mining processes required enormous amounts of labor. The miners and muckers went down the shafts and out into the drifts, crosscuts, and stopes to toil by the flickering light of the single candle attached to each man's cap. They had three candles for a ten-hour shift, and the miner who used them too fast sat out the last of his shift in total darkness. About 1910 the industry developed the sunshine lamp, which was little better, and in 1913 carbide lamps. Electric cap lamps were not introduced until 1934.[21] In timbering the mine, the miner had to build the square sets of rough timber posts 12 inches square and 8 feet long, or 5½ feet long if it were a horizontal piece.[22] This must have been exceedingly difficult in the half dark.

The miners began their day's work by first testing the overlying back or roof remaining after the previous day's blasting. This was done by prying loose any unstable rock found on the roof while hoping that the whole thing would not cave. The process was called barring down.[23] This finished, the actual work of mining began. The hard-rock ore is much too resistant to be removed by a pick. Whether a miner was breaking down ore, running a drift, or putting in a shaft, the method was much the same. Holes 12 feet long were drilled in the rock. When enough holes had been drilled in the proper manner, they were packed with explosive (black powder in the early days and 40% gelatin later) and detonated.[24] The actual blasting was done between shifts. The drilling itself was first done by hand. If the miner worked alone, he would hold the drill steel in one hand, rotating it constantly while striking it with a sledgehammer held in the other hand. This was known as single jacking. The Cornish despised this type of drilling and were known to have initiated strikes to avoid it.[25] More common was the method called double jacking, where one man held the drill steel and rotated it while his partner, using a two-handed grip, struck it with a sledgehammer.[26] Given the lighting and footing, the nerve of the man holding the steel must have been strong indeed.

The Homestake introduced the first mechanical rock drills in 1894, when fifty of them were issued.[27] They were not in general use until after the turn of the century. They were an improvement but extremely heavy and cumbersome. A central air compressor drove them. The drill steel was fastened to a piston, hitting a blow against the rock with each stroke of that piston and the steel reciprocating with each stroke.[28] The machines required two men to hold and operate them. All the drilling was dry until 1923, so the accumulated dust of hundreds of drills hung in the air for the men to breathe. There was no mechanically forced ventilation until 1923 to move the dust,[29] and silicosis was a constant danger.

The muckers assisted the miners in their work. These men, whose primary task was to move the broken ore into cars, also assisted in laying track and timbering. Their main tool was the shovel, and they had to be proficient with it. A mucker, when hired, was on probation. He was given three days in which to become capable enough to load 16 one-ton ore cars during a shift. The mucker who was unable to do this was fired, and another could try. Experienced muckers developed methods of cheating at the job by the judicious wedging of large blocks of ore in the car in a manner designed to leave considerable empty space. This was a great help unless the shift boss discovered the subterfuge. Some ore was too large for a mucker to handle. When this was the case it would have to be blockholed: a miner would drill holes in it, and it would be blasted. In some cases a blockholer could break rock for three muckers and in other cases only two.[30] The mucker worked hard for relatively low pay, but he was learning the trade and could work up to being a full-fledged miner.

The ore, blasted from bedrock and reduced to maneuverable size, had to be moved from the stopes through crosscuts, drifts, and tunnels to the shafts where it could be raised to the surface and thence carried to the mills. The first movers of ore from mine to mill were the teams of horses used by the Manuel brothers to haul ore to their arastra in the winter of 1876–77.[31] When the Hearst syndicate took over the property, horses still hauled from the open cut. In 1879, however, the company laid tracks and purchased a five-ton Baldwin locomotive—the J. B. Haggin by name. Two more steam locomotives, which were used until 1901 when a compressed air system was installed, augmented the Haggin. Each steam locomotive handled 30 tons of ore, while the compressed air locomotives could haul 32 four-ton cars, or 128 tons net.[32]

When the mine was first opened in depth by means of the Star and Vertical shafts, the ore from the stopes was close to those shafts and was trammeled by hand to the landing at the shaft. Later horses and mules were introduced—some 90 head of horses being underground at any given time.[33] Some horses were still in service in 1909 when the mine shut down. Many of them saw daylight for the first time when they were necessarily raised to the surface; they had to be taught to eat grass. A sling rigged beneath the cage raised and lowered the horses from one mine level to another. After the sling was affixed to the horse, men shoved the animal into the void of the shaft. The cage then took it to the desired level, where the miners hauled it onto solid footing again.[34] The Star shaft was used for these operations. Its cages did not handle men but transported the fodder for the horses as well as dynamite and candles.

The day of the horse lasted only until compressed air locomotion proved to be so successful on the surface that it was simply a matter of time until it took over underground. In 1904 the company installed a successful five-ton compressed air locomotive in the mine. The following year an Ingersoll-Rand compressor with a capacity of 1,800 cubic feet of free air per minute was installed and locomotives were purchased to replace the horses.[35] In order that the locomotives could be recharged with air, a separate line of high-pressure pipe was carried into the mine and to the end of each level.[36] In addition to hauling ore, these locomotives moved men in the mine, getting them to their stations at the beginning of work and taking them to the cages when their shift ended.

Under the original transportation system, the one-ton ore cars moved to the cages where they were hoisted to the surface and hand-trammed to the crushers located on the top floor of the mills. This utilized several shafts: the Star (the first sunk by the Homestake), the B & M (also known as the Old Abe), the Golden Prospect, Old Brig, Golden Gate, and the Ellison.[37] The shafts were originally separate units sufficient unto themselves, but with the passage of time they were interconnected with underground workings into a coordinated system. The depths of the shafts increased as the mining operations demanded. The Star, for example, went down 900 feet in 1900 and extended 1,250 feet in 1920. The Ellison was 600 feet in 1900 and 2,300 feet in 1928.[38] It pushed down still further in the 1930s.

The Ellison hoist, the pride of the company, had two components. The steam hoist located there originally carried a cage for men and a cage for ore cars. The second cage was replaced by a counterweight. In 1921 an electric skip[39] hoist and crushing system were put in the Ellison. The skip had a capacity of 4,000 tons in two shifts of fifteen hours from the 2,230-foot loading pocket. The skip could be removed rapidly and replaced by a double-decked cage to facilitate the moving of men at the changing of shifts.[40] The shaft, which originally contained three compartments, was enlarged to five: one 5 × 10-foot cage compartment, two skip compartments $5\frac{1}{2}$ feet × $5\frac{2}{3}$ feet, a pipe and ladder compartment, and a compartment for the cage counterweight and electric cables. The shaft was entirely timbered.[41] The primary crushing of the ore to $4\frac{1}{2}$ inches was conducted adjacent to the shaft at the 800-, 1,400-, and 2,000-foot levels by 36 × 48-inch jaw crushers driven by 125 horsepower, 2,200-volt electric motors.[42] Below each crusher station was a 1,500-ton ore bin and a loading pocket. These loading pockets were 75 feet below the levels on which the crushers were located. The ore was hoisted up from them in seven-ton capacity self-dumping skips. When the skips reached the surface, the ore automatically dumped into a bin and from there fed onto an apron conveyor which carried it to a revolving sieve (called a trommel) which sized the ore. The coarse ore moved onto secondary crushers and the fine went into a bin below.[43] At this stage the ore was ready to be loaded into cars and transported to the mills. The system used at the Ellison changed from time to time and varied somewhat from that used at the other shafts, but in the main the methods were similar.

The first ore coming from the Homestake mine was milled by a crude arastra, in turn replaced by a 10-stamp mill. When the Homestake Company was organized, Hearst shipped an 80-stamp mill to Lead, and from that time on, the pounding of the stamps was the trademark of the Homestake and Lead. In 1890, 240 stamps were dropping; in 1900, 800 were in operation; four years later 1,000 stamps were crushing ore. The height of the operations came in 1922, when the addition of the South Mill raised the total to 1,020 stamps. After 1922 the number of stamps dropped as ball-and-rod mills replaced them. The last stamp mills were retired in 1953.[44]

The process of stamp milling was similar in principle to the mortar and pestle used by a druggist.[45] Essentially a heavy body fell on the ore in a way to separate gold from the worthless mineral encasing it.[46] In the process a stamp dropped onto a die or iron plate inside a mortar, crushing the ore. Water containing the pulp or crushed ore surged out through a screened opening in the front of the mortar and spread over an inclined table lined with mercury-coated copper. The gold amalgamated with the mercury while the quartz passed on.[47] The amalgam, a combination of gold and mercury, was then refined, and pure gold was the result. This was the basic process by which the ore eventually released gold.

The original Homestake stamps weighed 880 pounds each and dropped 85–88 times per minute. The height of the drop was $9\frac{1}{2}$ inches, although this would vary as the dies wore down.[48] There were five stamps in each mortar. Each stamp crushed two tons of ore per working day in the early period.[49] Inside the mortar was the copper amalgamating plate onto which the mercury was fed in proportion to the richness of the ore.[50] The pulp, coming from the mortars and passing over the $4\frac{1}{2} \times$ 10-foot copper plate, finally moved into a launder (a trough or chute) below, which discharged it into the creek. The amalgam recovered was hand squeezed through a chamois skin and retorted in the boiler room of the mill.[51] Probably 70 percent of the gold was recovered;[52] the rest went down Gold Run Creek. Twice a month the mills stopped for cleanup. Workmen opened the mortars, took out all the dies, and removed the battery sand, which was panned or run through a rocker to recover the amalgam that was loose in the bottom of the mortar. They cleaned and scraped front and back plates with chisels. Nearly a day was necessary to clean up an 80-stamp mill.[53]

The key to the Homestake's success in combining ore-crushing and gold-saving capacities in their stamp mills lay in their mortar design, which became the standard of the mining world and was basically a modification of the design of the Union Iron Works of San Francisco.[54] The mortar was unusually narrow and allowed for quick discharge of the pulp, thus giving it nearly twice the capacity of the California batteries. The depth of the mortar permitted the amalgamation process to take place within it and prevented the scouring of the inside plate.[55] Many derided the process as hurried and wasteful, but the most eminent of mining experts stated, "The milling practice from a purely technical standpoint is insufficient and inadequate, but from a commercial aspect of the matter it commends on the whole, barring certain details, as eminently successful."[56] Since making money was the object of the whole operation, the managers were satisfied. Besides, constant technical improvement continued. The company tested and rejected a steam-driven stamp table which vibrated so admirably that the amalgam was shaken clear off the tables.[57] Other methods worked more satisfactorily. Increasing the size of the stamps to 1,550 pounds each and boosting capacity to 15 tons per stamp per day improved milling.[58] Greatly enlarged silvered tables replaced those of plain copper. Amalgamation captured approximately 72 percent of the gold in the ore.[59]

One of the Homestake's serious problems was the gold that got by the amalgamation process. From the early 1880s attempts were made to reduce this 30 percent loss in the tailings. In 1882 the Homestake erected a blanket house where the tailings were passed over tables covered with strips of Brussels carpets. The strips were removed every four hours, washed, and returned to duty.[60] Everyone knew that the concentrates recovered from the carpeting contained gold, but nobody knew how to process it until the Deadwood and Delaware Smelter began operations in Deadwood in 1888. This company indicated that it would buy the concentrates and duly process them. The Homestake sold the concentrates for about 50 percent of their gold value.[61]

The jig was another attempt at separating the gold from the tailings. This machine concentrated ore by the reciprocating action of a screen through water or by the pulsation of water through a screen. In 1895 the Homestake built a jig house which contained twelve Hartz jigs and later several Evans slime tables and a buddle or vat. For several years the jig house shipped 50 tons of concentrate daily to the smelter.[62] The concentrates assayed from $6 to $8 per ton, but the cost of smelting was high, and a considerable amount of gold was still escaping.[63]

The Homestake was not alone in seeking a solution to the recovery of the wasted gold. This was a matter of concern to the mining world in general; a 30 percent increase in production could mean life or death to any mine and would be vital in the supply of money to a world that was increasingly on the gold standard. The ultimate solution was in the field of chemistry and became known as the cyanide process. The basic patents were issued to J. S. McArthur and R. W. Forrest in Great Britain in 1887–88. They were first used in New Zealand in 1889 and shortly thereafter in South Africa. The first plant in the Black Hills was located at Deadwood in 1892 but was not very effective. The Homestake was not overly interested because its loss was less than some others'.[64] By 1898, however, the company gave a five-year contract to Charles W. Merrill, a highly regarded California expert in cyanidation, to work out a system for the Lead operation.[65] Merrill first arranged a careful testing on a working scale to determine the probable rate of extraction and the costs involved. In 1899 the Homestake erected a 60-ton experimental plant to treat the tailings from the Amicus Mill. It operated this plant for two summers, during which the 6-vat and 2-tank system proved the feasibility of the process.[66] In 1901 Sand Plant No. 1 began operation, recovering $40,000 monthly and establishing low cost records. The total take from the cyanide process in 1901 was $327,000.[67] A second plant went into operation in 1902. The combined amalgamation and cyanidation recovery was then 88 percent.[68] In 1906 Merrill designed a plant to treat the slime (the part of the ore that is clay-based), and the company built the plant in Deadwood. In 1908 a regrinding plant was added. The recovery rate reached 94 percent,[69] of which 13 percent was recovered by the cyanide process and 7 percent by the slime treatment.[70] When general collapse of older portions of the mine forced abandonment of the older mills and concentration in new mills (the so-called South Mills) in 1919, the sand floors of the old mills were run through the milling process for their considerable gold content.

Only the sands and slimes were left after amalgamation, and these were conveyed by pipelines to the cyanide plants. Here the system forced them through settling cones, the overflow or slime going to a slime thickener and the underflow going into the cyanide tanks. These tanks in Cyanide No. 1 were 10 feet 8 inches in depth, 44 feet in diameter, and held 700 tons of sand. They were built of redwood with heavy steel hoops for reinforcement.[71] A lime slurry was added to the sand as it passed into the tanks. The sand was deposited in the vat by a distributor in a process that took eight hours. When the tanks were filled and the sand had settled, the water drained and discharged into the creek.[72] Compressed air was forced through the sand for fourteen hours in order to increase the oxygen content sufficiently to dissolve the gold. A strong cyanide solution, applied for sixteen hours, released the gold. The solution then drained for fourteen hours. This process was repeated three times with progressively weaker cyanide solutions. In all steps zinc dust was added to the cyanide solution to precipitate the gold. The resulting solutions were pumped into filter presses where the zinc and gold remained while the solution drained off. The remaining precipitates contained all the gold and went to the refinery.[73] The sand waste was at first sent down the creek but later was used to backfill in the mine.

The process for extracting gold from the slime was very similar to that used on the sands. The slime went through pipes to the plant in Deadwood. Here it emptied into tanks from which it flowed into a series of filter presses. When the presses were filled, the pulp formed a compact cake on each frame. Cyanide, forced through the presses, dissolved the gold and thence was piped to cyanide solution tanks. Zinc dust was added as the slime went into precipitate presses. After the zinc and gold united and precipitated, the solution drained into tanks and was reused. The precipitate was then returned to Lead to the refinery. Work in the slime plant was very difficult. The presses were hollow frames with double canvas between each frame which required constant cleaning and changing.[74] In both the cyanide and the slime plants the work was exacting. Constant testing and watching, with little margin for error, was essential.

The final steps in the processes that turned the ore into gold occurred at the refinery. The crude gold came to the refinery in two forms—amalgam from the mills and precipitate from the cyanide and sand plants. The amalgam was actually crude bullion. To recover the pure gold it was necessary to remove the mercury that had been added. This was done in a retort furnace where the mercury vaporized at 1500° F., leaving a combination of gold and silver. The mercury leaving the furnace passed through cold water and returned to its original form to be reused.[75] The precipitate was treated by adding limestone and fluorspar. The resulting mixture was melted, and after the gold and silver had sunk, the rest was poured off as slag.[76] The bullion from both amalgam and precipitate was then remelted, and chlorine gas was forced through it, forming silver chloride, which floated atop the gold and was skimmed off. The silver chloride was treated with acid, yielding 85 percent pure silver. The gold was poured into bars 995 to 997 fine. The bars were cooled, weighed, stamped, recorded, and then shipped.[77] They went to New York in the early days, but later the Denver mint was the recipient.

The refinery underwent changes in technique like every other aspect of the Homestake experience. At first the operation was rather crude. By the 1930s, however, great care went into the processes in order to avoid losing any of the gold. The very air leaving the building was filtered, fumes from the furnace were subject to high voltage electricity to save specks of gold, floor sweepings were saved and processed. The clothes worn by the workers never left the refinery and when worn out were processed also. The men's showers were filtered. The water used to scrub the floors was treated.[78] Nothing was left to chance.

In addition to the basic procedures of mining and milling there were innumerable other related jobs. The large-scale timber operations required all the skills of the lumbering industry. There were electricians, machinists, teamsters, mechanics of all types, and even farmers who worked for the company. All these jobs and more were vital to the operation of the Homestake. All procedures were important; yet all were subordinate to the actual production of gold.

The Homestake was exceptional because it put together a system wherein all the processes of mining and milling were integrated into a complete whole. Everything that could be manufactured by the company more economically than it could be purchased was made in Lead. The company generated its own electricity, repaired its own engines, logged and milled its own timber, sharpened its own drill steel, and forged its own castings. A complete monolith is impossible in an industrial society based on specialization, but the Homestake Mining Company was as close to being one as was feasible. This structure, combined with constant experimentation and research, provided the means for keeping the mine in operation long after apparently richer lodes were exhausted. The practice of blocking out vast amounts of ore years in advance[79] meant that the worker could plan on a job that would last—a luxury almost unknown in gold mining. This concentration on the long-range operation of the mine was undoubtedly good business for the company. The profits and dividends went on in a foreseeable manner that gave the Homestake economic justification for investing in the welfare of its labor force.

CHAPTER FIVE

THE COMPANY TOWN

RODMAN W. PAUL has said, "Because of the nature of mining, which removes natural resources but cannot restore them,
mining towns and regions have rarely been stable."[1] He goes
further when he states, "If any one characteristic stood out in
common to the mining West, it was instability."[2] There can be
little doubt that Dr. Paul knows whereof he speaks; yet these
pronouncements are not really applicable to Lead, South Dakota[3]—the company town of the Homestake Mining Company.
Lead achieved and maintained a degree of stability unusual if
not unique in the West. This was partly due to the presence of
the great ore lode, which decades of intensive mining failed to
deplete. Of equal importance were the benevolent policies
applied by the company to the town and its citizens and the
political and social responsibility exhibited by these citizens,
who were mainly company employees. These factors combined
to create a city that developed civic pride and, with one notable
exception, preserved civic peace.

In the only study extant on company towns in the American West, James Brown Allen found certain characteristics that
were usual in most of them.[4] He felt that these towns developed
where single industries operated in isolated sections, and that
at first the company developed the minerals and only later the
town. The usual features included company-owned housing
and a company store that provided all goods to the employees
without competition.[5]

Lead, South Dakota, does not fit the pattern of Allen's
company town. It preceded the company in point of time and
was not the creature of the Homestake. Other strong mining
companies were based in Lead and thus denied the Homestake
total control of the town for a quarter-century. There was a
company store, but it was never designed to have exclusive con-

55

trol of retailing in the community and was never without strong competition. By the time the company had achieved anything approaching complete economic control of the community, the policies concerning the store, home ownership, and workers' welfare had been operative for years. The situation in Lead, as it developed through time, was generally acceptable to both the citizens and the company. There was no reason to turn to the more direct controls usual in company towns. The Homestake did run Lead, however, and thus the town qualifies as a company town.[6] The methods used were those of manipulation rather than naked force. There was usually room for divergent opinion. These factors, combined with unequaled tangible advantages, resulted in a company town that was uncommon in the mining West.

Lead predated the company by nearly a year and, like the Homestake, was a product of the Black Hills Gold Rush of 1876. The center of that rush moved from Custer City to Deadwood in the spring of 1876, and prospectors fanned out from Deadwood seeking gold. One of these prospectors, Thomas E. Carey, worked his way up Whitewood Creek and followed one of its branches, Gold Run Creek, to its source. Near this source in February 1876 he built a log cabin, the first building in what was to be Lead.[7] On Gold Run Carey found color, and as news of his discovery leaked out, others came to join him. The presence of these men combined with the presence of gold claims urgently necessitated some form of regulated property ownership. As a result, twelve men met on February 21, 1876, and organized the Summit Mining District.[8] Carey became the recorder and exercised the right to the discovery claim of 300 feet. The regulations of the Summit Mining District stated the methods for obtaining both quartz and placer claims, limited their size, and provided a method for transfer of title. The district charged a fee of $1.50 for all recordings and transfers.[9]

A mining district was a highly useful device for controlling property and performing certain functions of law and order, but it was a stopgap at best. It could not provide governmental functions with any permanence, even at the local level. As population increased, it became increasingly desirable that a town be started. In the enthusiasm and disorganization of the rush period, three settlements began in what would become the Lead city limits. The dates and order of the foundings are still in doubt.[10] In any case, "Smokey" Jones, a miner whose "hands and face were strangers to soap and water"[11] laid out a townsite with a pocket compass.[12] Because 1876 was the centennial year of the Republic, the town was named Washington. At approximately the same time (believed to be in February) P. A. Gushurst hired one Von Bodingen, a civil engineer, to plat a townsite adjacent to Washington on the east. Gushurst named his town Golden.[13] Later it was known as the lower Washington section of Lead.

On July 10, 1876, Richard B. Hughes and some traveling companions turned off Deadwood Gulch and headed up Poorman Gulch. Upon reaching its head they observed some men making a survey on a hill to the northeast. Although it was not apparent to them at the time, they were viewing the beginnings of Lead City.[14] The population of the Washington-Golden area was expanding, and a new town seemed to be in order. The name "Lead City" was chosen because of the number of leads[15] in the vicinity. The founding party met at the camp of "Smokey" Jones. Charles James was elected secretary of the city and James D. Coffin became the city recorder.[16]

Lead City was not really a legal entity. It did not come under the laws of Dakota Territory or the United States of America. It was on the great Sioux Reservation that encompassed all of South Dakota west of the Missouri River. The founders of the city had no ready-made legal structure to adopt. Thus they decided to allow the local mining law to cover all disputes.[17]

Mining codes did not cover all the contingencies that arise in establishing a city, even a ramshackle city on the frontier. They provided for democracy in a relatively pure state, however. The miners voted to have lots in the city measure 50×100 feet. The land available was limited, especially flat land; consequently, there were to be no alleys in the rear of lots. The claimant of a lot had to build on it within sixty days in order to hold his title. In actuality, the owners usually built 25-foot buildings and sold the remaining half of the lots. This rather niggardly custom accounted for the narrow frontages that were commonplace in Lead.[18]

By 1877 lode mining was becoming established, the population was expanding, and the limits of the town were proving insufficient. J. D. McIntyre hired a civil engineer named Hopkins to make a survey and plat a townsite.[19] The survey extended the limits of the city further up Gold Run and spread them up and over the tops of the surrounding mountains.[20] At about the same time the towns of Washington and Golden were incorporated into Lead City.

Upon its acquisition of the Black Hills area by treaty February 28, 1877, and subsequent acceptance of authority over the regions, the United States required owners of real property to acquire a patent from the federal government giving them title to the property owned by simple claim up to that time. A city obtained a patent to its townsite from the same source; for that reason, J. D. McIntyre requested Congress to issue a patent to the townsite of Lead.[21] At about the same time, miners discovered that the Homestake vein ran directly beneath the townsite for which Congress had authorized the patent. Federal law forbade the placing of townsites upon mineral-bearing land in order to prevent title disputes and to encourage mining. The Homestake Mining Company and other mineral producers had also asked for and received patents to their claims. The owners of property in Lead were extremely fearful that the Homestake might force them to vacate the mining surface. The company was afraid that its ability to follow the lode beneath the surface was jeopardized by the title invested in townsite owners.[22] A struggle was inevitable between the two opposing titleholders.

The legal contest between the townsite and mineral claimants took place in the federal courts, and the most that either side could claim was a draw. The mineral claimants, primarily the Homestake, brought suit in the United States District Court at Yankton to set aside the townsite. Whether they really wished the townsite set aside or merely wished some sort of permanent settlement that would not prevent continuation of the mining operations is still not clear. In any event, litigation brought the Homestake into possible trouble. The suit endangered the homes that belonged chiefly to its working force, and the workers were concerned at the threat to their domiciles.[23] The case hung fire for a number of years, always a point of discontent. The final agreement did not come until 1892 and was by stipulation rather than court decision, although there was no precedent for such a solution.[24] The court agreed to it, nevertheless, as did the General Land Office in Washington, D.C.[25]

According to the stipulated agreement, the mineral claimant, the Homestake, acquired the patents that carried title to the land and underlying minerals. In turn, the Homestake gave the surface occupants a contract that guaranteed their right to surface occupancy. The depth was not stated. George Hopkins resurveyed the townsite in order that the agreement be specific. Certain lots (A, B, C, D, and E on the Hopkins map) were exempted from the agreement and the owners received clear title to them. The mineral claimants agreed to pay for any damage caused by the mining operations to any property on the surface. In the event that the mining operations conducted in good faith required the removal of property on the surface, ninety days' notice was to be given.[26] The mineral claimants agreed to pay for any damage caused by the mining operations to the streets and alleys of Lead.[27] In effect, the Homestake was the landlord of the city.

The agreement provided for its own enforcement. Trustees were to be appointed to administer it. Leonard Gordon, the probate judge who received the patent from the government, was the government's trustee; Cyrus H. Enos, Ernest May, and P. A. Gushurst, prominent local citizens, were the trustees for the townsite claimants. The occupants of the townsite applied to the trustees for deeds. The Homestake was the major holder of the original mineral claims and eventually acquired them all. For all practical purposes the company owned the entire town. It bought nearly all of the property in the exempted lots. This property ownership gave the company a serious problem as well as enormous power. Regardless of who owned the property, the company's workers had to live on it. Some method of occupancy had to be worked out that would allow both occupancy and home ownership if stability of the working force was to be realized. The Homestake decided not to sell the property or to lease it; both of these methods could serve to alienate the mineral rights and jeopardize the mining operation. The dilemma was resolved by the issuance of licenses, revocable upon ninety days' notice, to occupy company-owned land.[28] The company legally had the right to evict anyone upon such notice and to force him to move his property at his own expense. This right if abused could have been used to enforce political and social conformity. But it was not abused. The privilege of revocation was used on rare occasions, and only when the company felt it to be of extreme necessity. On one occasion a licensee whose property blocked a proposed road to the South Mill refused to move or to sell. The Homestake revoked the license but paid for removal of the property and even paid damages for inconvenience caused.[29] On another occasion a license was revoked in order to remove a house of prostitution.[30] There may have been other isolated incidents, but the power inherent in the situation was largely unused.

Lead itself was always unusual among towns, even mining towns. In the beginning it was nearly inaccessible. A veritable wilderness of pine, spruce, underbrush, and fallen timber surrounded the embryo city. It could not be reached by way of Gold Run Creek.[31] The first merchant, P. A. Gushurst, packed his stock of groceries and dry goods in on his back, presumably making a number of trips.[32] By 1878, however, Louis Janin, a San Francisco mining engineer, visited the Black Hills and was impressed with Lead. He described it as a very superior mining community with ably administered justice and protection to life and property.[33] An eye-witness account of the city in the 1890s suggested: "You should go to the town of Lead, perched up in a narrow crooked gulch to the south and high above Deadwood, if you want to see what gold can do."[34]

Not only was Lead's appearance unusual but also its sound. The visitor was always able to hear Lead before he saw it. The hundreds of pounding stamps created a roar that was a trademark of the city as they slammed up and down seven days and nights a week.[35] The city itself, although surveyed and platted, was not built in a normal grid pattern. The configuration of the land, with its mountains and gulches, forced arrangements wherein streets followed the contours of the land. The streets branching off Main Street frequently stopped completely when the slope became too steep. Wooden staircases—regarded by the city as streets, named as such, and maintained by city crews—continued up the mountainsides. The houses in the town were mostly jerry-built, in the tradition of the mining camp which regards everything as essentially temporary. They were built on lots scratched out of the steep mountainsides, appearing to perch on top of each other in ever ascending rows. Many houses appeared perfectly normal on the street side, where one might walk in on the ground level, but have a sheer drop for a back yard. The streets were frequently too narrow for two vehicles to pass abreast—one would have to straddle the sidewalk. Most streets twisted and turned seeking a foothold on the mountains that allowed a feasible grade.

From its very beginnings the town gave the impression of an almost total lack of planning and orderliness.[36] Even the main thoroughfare had an odd look. The wooden sidewalks, which lasted until the 1920s, were not built in a continuous line but in a step-down design with steps connecting the continuous stretches in order to compensate for the steep grade.[37] The street itself followed the bottom of the gulch and the bends of Gold Run Creek. In later years the creek ran through underground pipes, but the streets still followed the original layout. The city never really changed its appearance from the early days. There were new sidewalks and storefronts but only a few new houses. It was always a mining camp.

The principal landmark in Lead was the Open Cut. This huge gash in the rock formations was the result of the early mining operations wherein the ore and waste rock were pulled into the mine from below. Its variegated walls showed the strata of the rocks in various hues of great beauty. As long as mining continued, it continually grew in size. The business district of Lead was originally built adjacent to the cut, as were various Homestake structures, the Star shaft, and the superintendent's home. The growth of the cut encroached upon these. In addition, the mining practices followed during the early years had not included backfilling, so the timber stopes in the upper levels caved when the timber rotted.[38] The net result was a general settling of the area that began in 1919 and lasted into the 1930s.[39] The intersection of Mill and Main streets, the center of the business area of the city, went down 28 feet in all. The Smead Hotel, one hundred rooms and the finest in the state, was ruined and torn down. Gone were the Northwestern Railroad depot, the First National Bank, the Smith Block, the Campbell Building, and the Ford Theater.[40]

The Homestake moved all its mills and relocated its shafts to the south of Main Street. The superintendent built a new house nearly a mile to the west. The business district moved up Main Street to the west, but much was gone for good. The Smead was never rebuilt. Many business firms moved to nearby Deadwood. The red-light district, which had been disguised as a dance-hall section, was also wiped out.[41] Fanny Hill's, the

Green Front, and the other houses of pleasure were as subject to the problems of settling as was the Episcopal Church. All had to go.

Nevertheless, Lead, unlike many mining towns, grew steadily and never experienced a sudden population drop as many camps did. The Homestake Mine, developed as a long-term proposition, was the major reason for this. Quartz mining had begun giving a broad base for employment in 1878, and the population was an estimated 1,500.[42] The *Lead City Daily Tribune* in 1882 complained that Superintendent McMaster's policy of forbidding the construction of houses near the Homestake works, apparently formulated in fear of fire hazard, was slowing the growth of Lead by preventing the erection of needed housing.[43] This was only a temporary stoppage, and the growth continued. By 1890 there were 2,581 people living in the city.[44] In 1900 the United States Census showed the city to have 6,210 people.[45] It was the largest city in the Black Hills by nearly 3,000 people. Lead leveled off at a population ranging from 7,000 to 10,000 thereafter. With the advent of tourism in the 1920s and 1930s it lost its edge in population to Rapid City. In a sense Lead was an anachronism. Its mining remained strong when industry in the surrounding region failed. It never developed any new industries and did not cater much to the tourist. When the rest of the Black Hills area was changing, Lead remained the same mining town.

The Homestake Mining Company's mines and mills provided the main reason for the town's existence. These, in addition to the sawmills, railroads, water systems, and land ownership, provided an economic base that was bound to extend to areas beyond those strictly connected with their work. The owners and managers of the Homestake had interests in other vital facets of the city as well. Chief among these was the "company store," the Hearst Mercantile Company. In the first years of the Homestake's founding, the company had been hampered by shortages of supplies, partially due to lack of proper retail outlets in Lead. As a result, Senator George Hearst started the store in 1879 through the offices of his personal representative in Lead, Thomas James. James organized and operated the store under his own name until 1882 when it assumed Hearst's name. Locally it was always known as the "brick store." Although it was originally begun as an enterprise designed to supply the Homestake, it was quite natural that its merchandise would be desired by the townspeople. It thus became a retail establishment and was self-supporting.[46]

The Hearst Mercantile carried practically everything. It distributed Laflin and Rand powder, Ingersoll machinery, and the products of nearly every other representative concern in the mining field. Because of its vast stocks and large capital, it also became a wholesale distributor.[47] It sold dry goods, carpet, draperies, millinery, ladies' ready-to-wear, clothing, furnishings, shoes, trunks and valises, groceries, buggies, automobiles, furs, plumbing supplies, and lumber. For a while it even had a drugstore.[48]

Biographers of the late William Randolph Hearst have criticized this store on the grounds that it was a "company store" in the tradition of such institutions in southern mill towns. Carlson and Bates state that "the Hearst Mercantile Company of Lead, South Dakota, is the company store from which the employees of Homestake are expected to do their buying. . ."[49] This charge cannot be corroborated. There was no requirement that employees buy at the Hearst Mercantile and no attempt to pay in company scrip redeemable there.[50] The store owed its undoubted success to two factors—the widest selection and greatest variety of any store in the region and a system of unlimited credit for Homestake employees. It was literally true that a man could enter the store with only a Homestake employee card to his name and come out with food, clothes, furniture, and a new car—plus a considerable bill. His forthcoming check would be sent to the store, an agreed amount taken out, and the remainder given to the wage earner. Some men were in debt to the store for an entire lifetime, but the credit was easy with no interest charges, and no one minded if anyone bought elsewhere, which was easily done.[51] The Lead newspapers showed business advertisements of firms carrying items available in the "brick store." In any case, a stranglehold on the market was unthinkable with Deadwood, a very active retail center, only three miles away, easily accessible by interurban railroad.[52]

The Homestake also had close connections with the banking business. The banking house of Thum, Lake and Company—an old gold rush bank—was reorganized into the Lead City Bank in 1883 and became a state bank in 1890.[53] It became the First National Bank, with assets of over $500,000, in 1891. T. J. Grier was the president; R. H. Driscoll, Grier's confidential agent, was cashier; and Dr. J. W. Freeman, the chief Homestake surgeon, was a member of the Board of Directors.[54] As evidenced by the officers, the First National was controlled by Homestake men and was popularly known as the Homestake Bank.[55] The relationship with the Homestake was on a personal rather than an official basis. This was natural. The company officials were the leading citizens of the town and solid men of capital and business. It was usual for such men to head banks in addition to their normal business. They served the bank well and enabled it to weather the storm of the Panic of 1893 by assessing each share of stock $40, which was added to the bank's reserves. The First National never closed its doors in 1893 or in any subsequent depression.[56] The bank expanded in later years and eventually had seven branches. The Lead bank was the chief bank in the system until 1938, when the headquarters moved to Rapid City.

The bank had more in common with the Homestake than just the same officers. Driscoll, the cashier and later president, did the actual purchasing of competing mining companies for the Homestake.[57] In addition, the bank served as paymaster for the company. It also served as the credit-extending institution for the Homestake in housing loans made to company employees. However, there were competing banks at all times, and company employees could do their banking business with them.[58]

Economic power leads inevitably to political power. There can be little doubt that the Homestake dominated Lawrence County, was decisive in the Black Hills area, powerful in the Second Congressional District, and strong in the State Legislature. The company could occasionally swing powerful weight in the Congress of the United States. The Homestake was well supplied with political talent from the beginning. McMaster engaged in politics; Gideon C. Moody, a master politico in territorial days, later became a United States Senator from the springboard of his position as Chief Counsel.[59] Moody's leadership in the Republican party not only marked the open emergence of the Homestake Mining Company in politics but also tied the majority of the citizens of Lead to the G.O.P.[60] Homestake power and prestige, combined with strong organization, meant a long series of defeats for the Democratic party in Lead.

Oddly enough, the Hearsts were Democrats in an active way. George was a United States Senator from California; William Randolph, his son, ran on several Democratic tickets and was a power in the party through 1932. Yet neither apparently made any effort to prevent the creation by their company of the strongest Republican machine in South Dakota.

The Homestake, following the example of Moody, continued to operate in politics through its legal department.[61] The Chief Counsel, first Moody and later Chambers Kellar, was always a power in state politics and practically invincible locally. In the period before the turn of the century, the power of the company was used blatantly. The shift bosses, very handy men in many ways, passed the word to the workers as to how they should vote. There is evidence that they went so far as to give marked sample ballots to the men in order that there would be no mistake.[62] There was no real way the directions could be enforced, but it cut down on vocal opposition. In the same period the immigrant workers were organized into racial and ethnic clubs. In the election of 1896 the so-called hyphenated clubs marched in torchlight parades for McKinley and gold. There were Afro-American, Slavonian-American, Italian-American, and similar other Republican clubs; each had its own distinctive uniform.[63] It worked well enough to beat Bryan in Lead. There were other groups, however, who favored the Democrats. The Irish, some Italians, the radical Finns, and a portion of the Slavonians voted Democratic. In the period from 1900 to 1910, a split on the lines of preference for either the company or the union appeared. Those favoring the company tended to vote Republican and those for the union voted Democratic.[64]

The company's main defeats on the local level came when it attempted to dictate voting on either men or issues when the mass of workers could see no reason for it. The workers usually went along with the company if they thought the issue was generally in the best interest of the Homestake and hence their jobs. If this were not the case and pressure from the company became irksome, they might very well decide to bolt.[65] As time passed, the methods became more subtle, but the result was the same. As the generations passed, the dedication to the G.O.P. increased, if anything. Even the Catholic voters, normally Democratic in most places, gave the Republicans at least an even split.

On the state level the Homestake mainly tried to protect itself from burdensome legislation. This was essentially a negative role but a necessary one. The company's only real defeat occurred in 1936, when a Democratic legislature passed South Dakota's first ore tax. The Homestake, making enormous profits at the time, had little to justify blocking the tax to a legislature of a dust bowl state racked by the depression.[66] Homestake influence was also evident in the appointment of the state mine inspector, who was always from Lead and essentially sympathetic to the company.

For all its power, the Homestake never controlled South Dakota in the sense that the Southern Pacific Railroad controlled California or Anaconda Copper controlled Montana. It made no effort to do so. The company did control Lead and Lawrence County as long as it practiced restraint and kept the mail fist well hidden beneath the velvet glove. Homestake influence permeated community affairs. The school board always had Homestake men on it, as did the city council. Yet, in a one-company town, who else could there be? Since company policy usually was very generous toward both the school and the community, friction was at a minimum.

The company's influence extended as time passed. From its beginnings the town had a strong middle class composed of store owners, mining speculators, teachers, and professional men. When the bulk of the retail merchants moved to Deadwood following the settling of the business district, this class decreased drastically. By the 1930s the number of independent businessmen and teachers was insignificant in relation to the total population. Lead had increasingly become a town composed of people who either ran the Homestake or worked for it. This caused no particular trouble between management and workers but tended somewhat to stifle the social structure.

Lead was physically superior to most of the mining towns of the American West.[67] There was no smelter to wipe out the sky and blight the grass and trees. The people in Lead had professional fire and police protection, paved streets, free water, a fully equipped hospital, modern schools, and excellent recreational facilities.[68] The Homestake's policies that fostered such advantages helped the people. They also helped the company by increasing the permanence of the labor force. By granting free occupancy to company land while at the same time providing sources of credit for home building, the Homestake helped to create a home-owning class of workers with a stake in the community and in their jobs. Such workers could not easily pick up and leave at the first sign of trouble or at a whim. This group proved to be the core of solid responsible citizens that any stable community must have. During the lockout the majority of these men and their families stayed in Lead and provided the continuity that allowed both the company and the community to rapidly return to normal.[69]

One of the main reasons for the instability of most western mining towns was the nature of mining itself. It removed the economic base of the communities without the slightest prospect of replacing it.[70] The size of the great Homestake ore body precluded this sort of instability in Lead. The ore was there to provide the jobs and the funds that allowed the welfare programs and the good physical conditions. Yet the quirks of nature were not the sole determinant. Butte, Montana—with its rich hill of copper—is one of the great examples of concentrated mineral wealth in the world; its story is one of poor physical conditions, continuous labor troubles, and a generally unsettled atmosphere. The element of sound management combined with a certain enlightened self-interest certainly played a vital role in Lead. The Homestake and the people of Lead had combined their talents to create a company town unlike other company towns. Its unusual qualities allowed it to achieve a stability unheard of in the mining West.

CHAPTER SIX

WELFARE: 1877-1910

THE HOMESTAKE employee never stood alone, dependent on his own resources and utterly defenseless in the face of public or private disasters that might afflict him. Even in the company's early period, sources of help for him and his family were available. Some of these sources were institutional; others sprang from private humanitarian impulses. The help was never direct, as it would be later, but it was effective in its realm. In a later period the government predominated, but the employer was preeminent. Above all, it should be understood that the Homestake worker had the sources of help when he needed them and that this assistance became more systematized as time went on.

Much of the aid the worker received until 1910 depended on his own participation and contributions. This was certainly true of the benefits received from the union. The Lead City Miners' Union was, in one sense at least, an organization of individuals combined for their mutual good. The union provided a variety of advantages for its members, not the least of which was a building for their use and available to others in the community. The original $6,000 structure built in 1880[1] was replaced by a $70,000 building in 1894.[2] Many of the best touring companies in the country featuring variety entertainment, legitimate theater, and even grand opera[3]—as well as local entertainment such as the Singer's Club performance of *The Mikado* in 1899[4]—appeared under union auspices. Thus the Lead City Miners' Union provided the cultural center for the mining community. The union was able to provide this and still retain financial solvency. It had a surplus in its treasury by the turn of the century.[5]

In addition to providing culture for the community, subsidization of a union newspaper[6] and care of members in time

This photo, taken October 8, 1876, is believed to be the first picture ever snapped of the community which later became a part of Lead. The building on the left is the St. James Hotel. The area is below the Homestake Substation still known as Washington.

Lead's Main Street in the late 1870s.

Lead about 1890.

Miners in the 1880s posed before an open ledge.
Double-jacking is illustrated here.

Sam McMaster, the first
superintendent of the
Homestake.

Phoebe Hearst, widow of
George Hearst, mother
of William Randolph Hearst,
philanthropist and major
stockholder in the Homestake.

T. J. Grier, McMaster's
successor, at the height of
his power.

OPEN CUT
HOMESTAKE MINING CO.
LEAD, SO. DAK.

BLACK LINE (DOTTED) REPRESENTS THE OUTLINE OF THE HILL IN 1876
BEFORE IT WAS MINED AWAY
APPROXIMATELY 20,000,000 TONS OF ROCK (ORE AND WASTE) TAKEN
FROM THIS CUT

METHOD OF MINING

DOTTED WHITE LINES REPRESENT DRAW HOLES THROUGH WHICH THE ROCK
RUNS BY GRAVITY TO THE LEVELS BELOW WHERE IT IS DRAWN OUT
OF CHUTES INTO ONE TON CARS. THE ORE IS HOISTED TO THE
SURFACE FOR TREATMENT IN STAMP MILLS. THE WASTE IS USED
FOR FILLING EMPTY OR WORKED OUT STOPES.

PHOTO & COPYRIGHT BY A. ROSE

Two shift bosses (identified by the large carbide lights they carry) inspect the timbering while a miner drills by candlelight.

Horse-drawn ore train. Some of these animals were born and lived out their lives underground.

Trainload of ore on the Caledonia 500-foot level, about 1890.

Cleanup at the Homestake stamp mills, 1888.

The new air-powered drilling machine, 1912. However, miners are still wearing soft hats and lighting is by candle. Each miner got three candles issued to him per shift.

Men waiting to go down in the Ellison double-deck cages, 1908. Most of these miners were engaged in the labor troubles of 1909–10.

An unusual shot, showing the mucking out of a large shrinkage stope at the 500-foot level of the Caledonia, 1926.

Square-set timbering in a shrinkage stope, about 1929.

An old-style gold shipment. The bars are much bigger than those used today and contained silver. The well-armed company guard was also a deputy sheriff.

Cyanide vats filling with sand and water, 1932.

A recent photograph showing the disintegrating timbering of older portions of the mine.

The company store. The Hearst Mercantile Company in the 1930s.

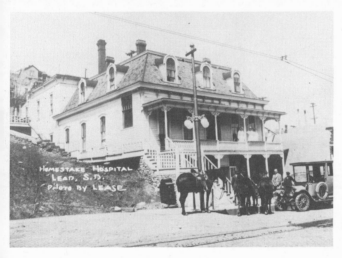

The Homestake hospital. This is the second building. It replaced a log cabin in 1889 and was in turn replaced by a brick building in 1923.

The recreation and library building and the Homestake theater. On the right is the bronze statue of T. J. Grier.

The Homestake today, showing the Yates shaft with most of the mills below it.

of illness, injury, or death show that the organization was carrying out its proper functions.[7] The union was very concerned with maintaining good relations and preventing misunderstandings with the company. It investigated any complaints and periodically inspected the company hospital to insure that its members received proper care.[8] It took care of the immediate and local functions of unionism and received the support of its members and the respect of the company in so doing.

With the organization of the Western Federation of Miners in 1893, the union took on other functions. It became the link connecting local members with labor as a whole and providing them with a sense of being part of something larger than themselves. The term used was "solidarity." It was exemplified through generous aid to other W.F.M. unions in the form of strike funds and to the W.F.M. officers in their clashes with the law.[9]

The union acted in the interest of the worker as an intermediary and a buffer in his relations with the Homestake. While it did not directly serve nonunion members, these men usually received such benefits as the union obtained. The Homestake worker did not need the active protection of the union since there was no real labor trouble prior to 1909. However, the fact of the existence of the Lead City Miners' Union served as a counterpoise to the company and as such benefited all.

If the Homestake employee were an immigrant or belonged to a distinct racial-ethnic group, he would probably receive support from members of the same group or from organizations to which they belonged. Nearly all the national groups had lodges. The Irish had the Clanna Gael and the Scots the Order of Scottish Clans,[10] both primarily social in nature. Among the bodies which went further and provided insurance programs as well as other benefits were the Sons of Saint George for the Cornish and the Society of Christopher Columbo for the Italians.[11] The Slavonians belonged to either of two societies, the Hrvatska Zajednica or the Broinhs Velebit. Swedish and other Scandinavian groups also provided the combination of insurance and brotherhood.[12] In addition to these advantages, the societies—while offering an opportunity to participate in choruses and play in brass bands—above all provided a chance to be united with one's own, to be part of something familiar.

The benefits of belonging to a national group were not limited to membership in a lodge, playing a cornet, being psychologically adjusted or even buried. The worker, upon coming to Lead, was met by fellow countrymen who would find him a place to live and board, introduce him to other members of the group, take him to the saloon that catered to his nationality, and start him on the necessary steps that eventually led to employment.[13] In case employment were not immediate, his fellows would give or lend him money to enable him to survive until hired. They would arrange for his credit at one of the many boardinghouses that catered to men in his position.[14] They might even find him a bride. The Slavonians in particular tended to send back to the old country for wives, a complicated operation requiring the service of a marriage broker to negotiate the dowry.[15] Group contacts were most advantageous at such times. When the marriage took place, the guests would start the couple in life with contributions in the form of gold coins stuck in an apple passed among the wedding guests. On occasions, $2,000–$3,000 were collected in this way.[16] The married couple would very likely establish their home in an area where others of their nationality lived. These areas of national settlement were not ghettoes. No one had to live there, and anyone could leave at any time. The people did so as a matter of preference.[17]

Perhaps the greatest impulse to aiding the workers of the Homestake came from Mrs. Phoebe Hearst. Quiet, gentle, and refined, she loved art and charity and was genuinely devoted to humanity.[18] Upon the death of Senator Hearst in 1891 she became the majority owner of the Homestake, and her influence increased. She did not decide company policy, leaving that in the hands of the superintendents, but she took the most active interest possible in the Homestake workers and the city of Lead.[19] Her son, William Randolph Hearst, often visited Lead with his parents and even managed a baseball team there. He was not really interested in mining, although he enjoyed the profits of the mine. During his political period he was wont to come to Lead to show his friendship with the working class in general.[20] His mother, always concerned with the city and the miners, used her wealth and influence to help them. She did this in a modest and unobtrusive way and avoided the smugness and self-righteousness typical of many rich philanthropists. As a result, the workers and their families admired and even loved her.

One of the obvious needs of the people of Lead was a library. Mining camps are not the highest type of urban culture at best, but a library could help toward that end. With the advent of the new immigration, large numbers of foreign-born residents—unable to read English and of insufficient means to maintain their own libraries—required somewhat specialized library collections. Mrs. Hearst resolved these needs when she gave a complete library to the community of Lead as a Christmas present in 1894. The largest audience ever gathered in the area at that time was in voluntary attendance at the Miners' Union Opera House to hear Superintendent Grier present the gift on behalf of Mrs. Hearst. Mayor L. P. Henkins accepted the gift for the people.[21]

The library was first installed in the Miners' Union Hall with Mrs. Ferris (the future Mrs. T. J. Grier) as librarian. It was used primarily as a reading room for the first two years; facilities were inadequate for checking out books. In 1896 the library moved to quarters above the Hearst Mercantile Company store, and books were then checked out.[22] The physical plant was extremely modern, attractive, and of the best quality. There were tables, chairs, a piano, and all possible comforts.[23]

The Hearst Free Library owned over 4,000 volumes of literature, history, biography, science, art, poetry, and fiction.[24] It also had an unusually complete collection of periodicals. By 1897 the library subscribed to 81 different magazines and newspapers. A large proportion of these were in foreign languages—Italian, Slavonian, Finnish, German, Swedish, Spanish, Lithuanian, Norwegian, and French.[25] The library also purchased books in these languages. Some were classics of the particular country; others were translations of American works. Mrs. Hearst showed serious regard for the plight of the foreign-born and wanted them to mingle with the rest of the community. The library helped serve this purpose.

Mrs. Hearst also desired that the library foster the finer things in life. In her donation she specified that musicales be presented in the library. Every month the strains of chamber music wafted among the books and magazines and the high notes of the sopranos rattled the crystal chandeliers.[26] The entertainments were so popular that tickets were issued—free of course—in order to hold down the crowds. The library also served as the meeting place for many of the clubs and societies of Lead.[27] In its capacity as a library, it averaged 150 people in attendance daily,[28] a remarkable figure.

Mrs. Hearst, in continuing to support the library, gave her permission in 1902 for the purchase of 1,000 new books to meet all the demands of study as well as amusement.[29] She maintained this support until her death.[30]

One of Phoebe Hearst's deepest interests was in education. She exhibited her concern for higher education by giving generous endowments to the University of California, but her deepest concern was for the kindergarten. In Lead the kindergarten was not a function of the public schools. On June 11, 1900, the Woman's Club started the first kindergarten in Lead. A teacher and three assistants taught in the basement of the Episcopal Church.[31] The kindergarten even sponsored mothers' parties, complete with entertainment for children while the mothers listened to inspiring lectures. However, the Woman's Club could not continue to support the kindergarten, so it planned that a charge of fifty cents a week per child would be levied to make the school self-supporting.[32] At this point Mrs. Hearst came to the rescue and furnished the money to finish out the year. Thereafter she was the sole support of the kindergarten and continued to be until her death when the school board took charge.[33] She took a great personal interest in the kindergarten and often visited it and brought books and music for the children.[34]

The kindergarten expanded rapidly under the loving and generous care of Mrs. Hearst. By 1901 three teachers taught 140 children, and more awaited the arrival of additional furnishings.[35] By 1904 four teachers worked full time at the kindergarten. All were graduates of the Phoebe A. Hearst Kindergarten Training School in Washington, D.C.[36] They taught children from the ages of three to six five days a week.[37] The school was well equipped and had a small private park for outdoor play.[38] The kindergarten was intended to aid the children of the company's employees and of anyone else in Lead. It gave those of foreign birth a start in learning and using the

English language and taught them American ways and manners.[39] It enabled these children to enter the regular school system more nearly ready to compete with American-born children of the same age, especially in the area of reading readiness. Yet it did more. Mrs. Hearst was as interested in the mothers as in the children. The kindergarten, by taking the children out of the home, gave the overworked mothers a few hours in the day to go downtown and mingle with other people. This was intended as a step in their Americanization. The children themselves brought a new language and new customs into the home from the school and also helped the process.

The charities of Mrs. Hearst went even further. She helped the public schools by buying construction bonds at a low interest rate. She made it a practice to give each and every church in Lead $200 a year to further its work.[40] She did many private charitable acts that were never noticed publicly, and she did them without condescension. She asked nothing in return, but she received affection, love, and respect. Even at the height of labor troubles that would come, there was no criticism of Phoebe Hearst. The most violent of those who hated the Homestake never said a word against her and probably would not have dared to do so.[41]

Phoebe Hearst could well afford her charitable impulses. Senator Hearst, on his death in 1891, left his entire estate to her. This estate was valued at $18 million, but the figure was deceptive. Most of the estate was in land and mining properties, such as the Homestake and Anaconda, which were still in a state of development and whose future earnings would make the estate of much greater value.[42] As the properties developed, Phoebe used the money for a variety of noncharitable purposes. She traveled widely, collected art, and lived well. Much of the income of the estate was used in financing her son, William Randolph Hearst. In 1895 she sold her seven-sixteenths interest in Anaconda to finance him in his invasion of New York.[43] Yet the fact of her wealth and the fact that she did not give it all away do not deny her philanthropic motives. A great many of those who had great wealth, including Montana's copper kings, had no thought whatsoever for the workers. They reaped labor troubles in consequence. There is no evidence that Mrs. Hearst gave in a calculating way in order to make money by her charity. Her whole make-up ran counter to this. Those who did give for such reasons lacked the personal interest in individuals that was constantly expressed by Phoebe. The little presents for the children show the person more than do the large bequests. When she died in 1919, she left an estate of $11 million, mainly to her son. She had given away $21 million in her lifetime.[44]

There was little that gold miners and their families could do to show their appreciation to one of the wealthiest women in the world, but they certainly tried. Her every visit to Lead was covered minutely by the local newspaper. In 1903 the men, women, and children of the city contributed nickels and dimes toward the manufacture and purchase of a silver loving cup that was sent to her. She acknowledged the gift in a letter that stressed her appreciation and promised them a visit when her health improved to tell them "what is in my heart for your loving and generous tribute."[45] She fulfilled her promise in 1905. During her visit she inspected the mines, even going underground. She visited the kindergarten and again brought gifts to the children. On November 8 a vast reception was held for her at Superintendent Grier's home. Business closed down on Main Street. The mines closed in order that the men might meet her. The twenty-year veterans of employment in the Homestake marched in a body to see her. In all, over 3,000 people attended the reception.[46] It should be remembered that hard-rock miners are not the most sentimental of men. They petitioned Grier for the right to march; the tribute was undoubtedly from the heart.

The philanthropic image of Phoebe Hearst carried over somewhat to the company she owned. There is little doubt that her influence and example in the better treatment of the employees affected company policy. The company, however, also had motivations beyond the philanthropic. For one thing, some of the welfare activities were forced by circumstances not really conceived by the Homestake. For another, these practices proved to be good business in the long run—the basis of the entire company policy. Thus philanthropy and self-interest coincided; moneymaking and good works were frequently the same.

The finest thing the Homestake could do for its employees was to provide them with stable jobs at good pay under reasonable working conditions. The company went beyond this and pioneered in areas virtually unknown to other mining concerns, such as the matter of housing for the workers. The company acquired practically all the land in Lead for its own purposes.[47] By so doing, the Homestake made it unnecessary for anyone living in its town to purchase a building lot, which was a substantial saving to a home builder. A further saving came from the Homestake's payment of all real property taxes on this land, leaving the homeowner only the taxes on the house itself. In a pragmatic sense, the price paid by the homeowner for these savings was a considerable one. He had no title to the land on which his house stood and could be forced to move at any time. This in itself made very little difference. However, the lack of title, or even of lease, made it difficult if not impossible to borrow money for home financing through normal commercial channels. On the other hand, the company wanted as large a percentage as possible of its workers to own their own homes.[48] The advantages of this were obvious. Homes meant a serious investment by hard-working people with a stake in their jobs and in the community—stable citizens. As Superintendent Grier put it, the city was composed of "responsible, decent people, who are breadwinners, and not loafers."[49] Such workers were not likely to become enamoured with ideas that would disturb labor relations and thus disrupt production. The Homestake itself never attempted to provide permanent housing for its workers. In times of stress it might provide something temporary; the Hearst Mercantile ran a hotel for single workers at one time. However, the company did want a homeowning class of workers, and this was unlikely to be a sizable group unless the workers had a source of long-term credit. Therefore, the company advanced the purchase price of the home and allowed the employee to pay either in deferred payments or by monthly installments.[50] The Homestake thus obtained the type of employee it desired while keeping the sole ownership of the land and the mineral rights.

The Homestake Mining Company was unquestionably one of the pioneers in the United States in the field of industrial health services. These health facilities began in 1878 when the company contracted with Dr. D. K. Dickinson to furnish medical and surgical services to the employees.[51] Dr. Dickinson, a graduate of St. Louis Medical College in 1876, had worked as a surgeon for the Father DeSmet, Portland, and Welcome mines.[52] His assistant was Dr. J. W. Freeman, a noted bone surgeon and the man who would succeed Dickinson.[53] In the spring of 1879 the employees paid into a hospital fund for the first time. The money was turned over to Sister Mary Edward who conducted a small Catholic hospital in a building on South Wall Street.[54] Later in the year the Homestake established its first company hospital, a four-room log cabin wherein one untrained male attendant administered to the patients. This structure was used until 1886, when it was replaced by a two-story frame structure. This building was enlarged in 1895 by excavating the dirt from beneath the building and thus adding a third floor.[55] The hospitalization was not free; the employees continued monthly payments of $1.10. For this amount the employee received both hospital and surgical care for himself and his family.[56] Drug and sundries charges were additional. The doctors also answered house calls, traveling by horse and buggy. Because of lack of space, obstetric cases were not handled at the Homestake Hospital; an arrangement was made with the hospital in Deadwood for care of these.[57] A Homestake doctor also normally served as the city health officer.

The contract system was not entirely satisfactory. Under it the company engaged the chief surgeon, who employed his assistants. The company had little control over the surgeon and few checks on the system as a whole. Employee dissatisfaction was also evident at times. There were frequent disputes over charges for borderline cases. The men were also disturbed because they were paying money into a fund over which they had little or no management.[58] This friction was more irritating than serious, but a phase of company activity that should have contributed to a harmonious relationship between worker and management was proving to be otherwise. As a result the company reorganized the system in 1906, putting the health service fully and firmly under the control of the company. The Homestake hired the chief surgeon and all his assistants plus all the nursing and administrative help. The doctors were no longer allowed to maintain private practices. The charge of $1.10 per month continued, but it was the only charge rendered and covered everything for both the worker and his family.[59] Employees were not bound to go to company doctors and could, if they desired, refuse to pay the $1.10 and provide their own medical care.[60] The reorganization, while helpful, did not entirely do away with criticism. The workers still had little voice in the management of the funds collected, and the monthly payments galled them. Criticism would continue until the program was absolutely free.[61]

The Homestake, as well as Mrs. Hearst, was vitally interested in education. The schools in Lead (kindergarten excepted) were supported by public taxation of course. However, to a great degree this support came from the company. The school taxes in South Dakota were almost exclusively taxes on real property. Since the Homestake paid the only real estate taxes in Lead, it became the sole support of the schools. Paying taxes alone is of course no particular evidence of concern for the public welfare or even of civic virtue. Yet the company with its stranglehold on local politics could very well have chosen to pay much less than it did. Actually, the company was consistently generous toward the school.[62]

Education in Lead started in 1877 when a Miss Graham opened a private tuition school. The first public school began later the same year in a log cabin on Gold Run Creek.[63] The school system grew with the community. Lead schools were organized under the district township system until 1896 when the Lead Independent School District #6 was organized. This district included an $8\frac{1}{2} \times 2$-mile area and was situated on some of the finest gold-bearing ore in the world, which provided a marvelous tax base.[64]

The interest of the company in the school district can best be illustrated by the case of R. H. Driscoll. Driscoll was a Harvard graduate of the class of 1881. He was employed by Marvin Hughitt, president of the Chicago and Northwestern Railroad, as his private secretary; in that capacity he accompanied Hughitt on a business trip to the Black Hills. While in the Hills he met and apparently greatly impressed Samuel McMaster. McMaster persuaded Driscoll to remain in Lead to supervise the public schools, and Driscoll accepted the challenge.[65] Evidently the school board raised no objection. Driscoll moved on to other tasks and eventually became the long-term head of the First National Bank. The company backed him always. Driscoll illustrated not only the company's control of the schools but also its desire for excellence in them.

The company, again wanting a stabilizing influence on the working man, felt that the family man would be attracted to employment as a result of the educational advantages in Lead.[66] The Homestake made sure that its influence for good education would always be felt. Company officials, always on the school board, supported progressive measures in education and certainly were not niggardly with tax money.[67] The Homestake encouraged scholarship and high standards. In 1899 Superintendent Grier donated $780 to be divided as competitive prizes for students of the grade and high schools.[68] Such actions go beyond the minimum necessary to maintain stability in the labor force.

Judging by external features, the schools were superior. The buildings were the finest in the state.[69] By 1897 the high school offered both an English and a Latin course.[70] The teachers were very highly paid. The superintendent received $150 per month and the teachers averaged $67.50.[71] This occurred during a period when teachers in the larger cities of eastern South Dakota received an average of $33 per month.[72] The school term lasted ten months, and teachers' qualifications were well regulated. A first-grade certificate was necessary in order to be employed, and periodical examinations were taken in order to hold a position.[73] By 1898 all teachers belonged to the Lawrence County Teacher's Association.[74] In 1906 the North Central Association of Colleges and Secondary Schools gave the Lead schools its unanimous endorsement. Only three or four schools in South Dakota and six in Nebraska were so recognized.[75]

The Homestake's concern with the public schools was both effective and in the public interest. The managers of the company sent their own children to these schools; yet it was not a selfish concern with their own that prompted the policy. The schools stressed manual training, domestic science, and other activities unnecessary to the members of the managerial class. The organization of night schools for adults in 1906 was another example of the school's attempt to serve the working man. This program charged tuition at first but later was free. Any course for which there was a demand was willingly offered.[76] Americanization classes were also held in the public school system in order to teach English to the immigrants and to prepare them for the examinations necessary for naturalization.[77] The company provided the district with lists of workers who were not citizens and strongly encouraged men to start the processes necessary to citizenship.

The essence of the Homestake's welfare policies before 1909 lay not only in the large-scale programs it supported but also in a general attitude of helpfulness and good neighborliness. The company was almost always willing to donate to any cause that was at all worthy. The management willingly listened to requests for time off, whether to greet Mrs. Hearst or to go to a ball sponsored by some national group. The Homestake was highly civic-minded. In 1902, for example, the company gave the city a strip of land that doubled the city's park system.[78] When a church had trouble with its physical plant, it was likely that a Homestake crew would find whatever was causing the problem and repair it, free of charge. If the lowliest mucker had a bright child, he would be encouraged to continue his education and perhaps be hired by the company upon graduation.[79] The company, although as large as any in its field, fostered personal relationships between worker and management which in themselves were valuable.

By the end of the 1877–1910 period the Homestake employees were well served by their company, by their community, by individual philanthropy, and by organizations born of their own impulse. They were well paid, had decent working conditions, and good job expectancy. They had a good opportunity to own their own homes and experienced little or no racial prejudice. Their lot was superior to that of most American working men and infinitely superior to most gold miners. It might have been expected that peace and tranquility would continue. Yet it was at this moment that an overanxious union and a hypersensitive management broke completely and plunged the Homestake and its workers into their only period of serious industrial strife.

CHAPTER SEVEN

THE LOCKOUT

THE Western Federation of Miners (W.F.M.) was in the process of change in 1909. This change was primarily in philosophy and would extend itself to personalities and tactics. At the heart of the change was the leadership of Charles Moyer, who was deliberately leading the W.F.M. away from the anarcho-syndicalistic Industrial Workers of the World (I.W.W.) and toward possible reaffiliation with the American Federation of Labor (A.F. of L.), which emphasized the immediate goals of higher wages, shorter hours, and better working conditions.

Moyer doubtless had many reasons for attempting this change. Bill Haywood of the I.W.W. thought that the experience of the long months in prison during the Boise murder trial had sapped much of the enthusiasm of the W.F.M. president.[1] While this is not unlikely, it is more probable that Moyer had reached several conclusions of his own. It must have been apparent that syndicalism had alienated American public opinion and was in fact a philosophy at odds with the America of that day. Tactically it had been disastrous. Violence more often than not brought upon the heads of the W.F.M. members revulsion, arrest, and federal intervention. Thus Moyer, by spurning the I.W.W. and gently wooing the A.F. of L., was trying to pursue a path more likely to bring success. In doing so, however, he had partially split the W.F.M. While he was able to keep the majority satisfied, there was still the hard core of syndicalists whose ideas had changed very little. They were continually sniping at Moyer and his lieutenants. Thus the internally divided W.F.M. was in the midst of an unfinished experiment when its next major test came unexpectedly at Lead, South Dakota. The opponent was the Homestake Mining Company.

In February 1909 William E. Tracy, a W.F.M. organizer, visited Lead to inspect and report on union activity. He reported Lead Union No. 2 as growing rapidly and being free of debt, although the Black Hills in general was not too prosperous.[2] At about the same time, the W.F.M. locals in the northern Black Hills asked the national organization for an organizer to help them in their efforts to unionize completely the various camps at Lead, Terry, Terraville, Deadwood, and other places. Tracy returned to the work on September 1, 1909.

Since the founding of the camp, the biggest event in Lead had been Labor Day. The union traditionally had charge of the program, and in the midst of its membership drive, the members outdid themselves on September 6. The parade was led by mounted police and featured twenty-five different Black Hills unions. The last in marching order, and the largest, was the Lead Miners' Union. The featured attraction of the afternoon program (preceding the baseball game and the acrobats) was "Mother" Jones, a seventy-five-year-old lady with silver hair and a sweet smile, who prided herself on being the matriarch of American socialism. She attacked capitalism strongly, saying, "We will go to Wall Street and say to the infernal crowd: 'Get the hell out of here.' No doubt I shock the ears of some pious people. But I am not on my knees praying to God. I am fighting God and if He don't like my way I tell him 'Mind your own business.' I am tending to mine."[3] Having thus injected the mellowing influence of woman into the community and having punched a would-be photographer in the eye, this sweet old lady left. She was perhaps a symbol of what was happening to labor in Lead as the hard-core organizers of the W.F.M. started to drift into town.

William Tracy—a crude, immoral, drunken man—was a surprisingly effective organizer. Working with a number of local people, he drew large numbers of men into the union. As his work progressed, members of the W.F.M. representing the various nationalities in Lead arrived to assist him. Yanco Terzich, a member of the Executive Board of the W.F.M. took charge of the Slavonians. A Mr. Lowney worked with Irishmen, while a Mr. Davidson had charge of recruiting the Cornish. These men had instructions to use all legitimate means available to increase membership.[4] The drive continued. Mrs. Emma F. Langdon of the *Miners' Magazine* addressed a mass meeting of the miners on October 11 and demanded that all former members of the union make up their arrears and become active. She stated, "Anyone who ignores this will be dealt with as the Union shall determine."[5] By October 24, 1909, the union claimed that 98 percent of the eligible Homestake employees were members of the union.[6] James Kirwan, Executive Board member from Terry, South Dakota, said there were only six eligible nonunion men in Lead.[7] Tracy and his associates had recruited nearly 1,000 men in a two-month period.

In the organizing process the union never approached the company officials. It felt that Grier had no objection to the union since he had shown none in the past. It made no demands on the company. The W.F.M. did not believe at that time in the check-off agreement, wherein the company would enforce union membership. It intended to achieve the closed shop in the Homestake Mining Company without touching the company management at all. Because the W.F.M. asked no positive action of the company, it fully expected that the company would not react to the major change in the status of its laboring force. But the W.F.M. was basing its tactics on an essentially false premise, because the Homestake could never agree to this change.

The campaign to sign up the last of the Homestake employees continued. A feature story in the Deadwood socialistic paper, *The Lantern,* attacked all who failed to join the union.[8] A series of mass meetings tried to whip up enthusiasm. The most important of these was on October 24. With Vice-President Mahoney of the W.F.M. looking on, William Tracy led the 300 men present to pass a resolution that union members would not work with nonunion men after November 25, 1909.[9] There are those who say that Tracy stampeded the meeting and that the majority of union members were dumbfounded by the action.[10] If this were true, it made little difference as events began to move faster than the efforts of anyone to control them. The next day a union committee asked Grier for a list of Homestake workers. Grier refused to give the list but told the men he was not opposed to unionism as such.[11]

Grier had been considering the union's moves for some time. Nonunion employees had complained to him of being intimidated by Tracy and his assistants to force them to join the union.[12] Grier also felt the influence of the Chief Counsel of the Homestake Mining Company, Chambers Kellar. Kellar was a soft-spoken, kindly man, who could be hard as a diamond when crossed. Grier may have sometimes wavered; Kellar never did. Kellar was always at Grier's side, urging him to take a definite stand against the union. The result was a lawsuit. The Homestake Mining Company sued the Western Federation of Miners for $10,000 damages. The stated causes of action were the union's interfering with the men who desired to work, driving them from work, and preventing them from going to work.[13] The lawsuit struck the rank and file union members with a severe shock. Some conservative union men told Tracy that the resolution must be withdrawn.[14] Tracy, however, said the union would import a lawyer from Denver and would contest the suit. *The Lantern* called the suit a bluff and attacked Grier as a hypocrite who posed as the friend of labor in order to destroy it.[15] The *Miners' Magazine* called the suit part of capitalism's concerted campaign against organized labor and suspected that it might be a stock-jobbing scheme to enable large stockholders to buy out the small stockholders.[16] This latter theory was untenable because Homestake stock, although listed on the exchange, was traded very little. It paid much too well for anyone to sell it. Besides, the syndicate had such complete control that there were not enough small stockholders to bother with. The union of course claimed that it had not used coercion but had used gentlemanly language in appealing to the manhood of the workers.[17] In truth, there may have been some coercion, but not on any large scale. There was social pressure, however. *The Lantern* was essentially right in calling the suit a bluff. The company never framed the issues of the case nor brought it to trial.[18] Because of the pressure of future events, both sides forgot the lawsuit.

The suit had taken the initiative from the union; its leadership was never able to regain it. From this point on, the W.F.M. reacted to counter company moves, and the next one came quickly. The following notice appeared in the *Lead Daily Call* on November 17, 1909:

Notice is hereby given that the Homestake Mining Company will employ only non-union men after January 1st, 1910.

The present scale of wages and the eight (8) hour shift will be maintained. All employees who desire to remain in the company's service must register at the general office of the company on or before December 15, 1909.

T. J. GRIER
Superintendent[19]

Grier's notice caught the union leadership completely unprepared. They apparently never thought that the benevolent Homestake Mining Company and its pleasant superintendent would turn on them in such a fashion. The union held a mass meeting that night, called to order by a local saloonkeeper and city alderman named John Mayo and chaired by William Tracy. The assembled body, voting as citizens of Lawrence County rather than as members of the W.F.M., approved a resolution "condemning the action of the Homestake Mining Company."[20] Copies of the resolution were sent to Mrs. Phoebe Hearst, William Randolph Hearst, James Ben Ali Haggin, and the directors of the company.

The drama came when Grier addressed the meeting. He said that he could prove the coercion of his workers and described the union, once benevolent and charitable, as changed and radical. He stated, "Men come to me and ask me if they must join at the end of a rope."[21] He accused union committees of entering the company hoists without permission and threatening workers with violence. Grier never finished his statement. Freeman Knowles—editor, lawyer, ex-congressman, and rampant socialist—interrupted him and harangued the crowd, insulting Grier personally. Knowles also claimed that he heard the mine was going to close and ended with the general query, "A hell of a system, isn't it?"[22] Grier left as Tracy vainly tried to quiet the crowd.

James Kirwan, the coolest and most sensible of the union leaders, telegraphed Haggin at San Francisco and asked him, as chairman of the Board, to rescind Grier's notice. Haggin replied that the matter was in the hands of Grier, in whom he had implicit faith. A committee of two men, Richard Bunney and Pete Jurey, was sent to Pleasanton, California, to see Mrs. Hearst, the principal stockholder. Mrs. Hearst received the men on her silk marquisette screened front porch and was extremely gracious to them. She explained that Mr. Grier had full authority and that she was sure he would do the right thing. On November 21 a committee composed of Thomas Ryan, James Kirwan, Pete Jurey, and Chris Christianson met with Grier and asked him to withdraw the nonunion order.[23] Grier, with the support of Haggin and Mrs. Hearst, had placed the union in an untenable position. The union never wanted a strike. Most of its members opposed a strike.[24] The Homestake had maneuvered the union into a position where a strike was the only alternative to utter destruction. Having thus outmaneuvered the union leadership, the company would soon move to deny the union even this last resort.

The union leadership, including Tracy and Kirwan, moved quickly after Grier's final refusal to withdraw the nonunion order. The national W.F.M. had given no permission to the Lead local to strike; without such permission, a strike could only be called with three-fourths majority vote of the local's membership.[25] Tracy hastily called a meeting on November 23, 1909, attended by perhaps one-third of the union membership.[26] Tom Ryan, secretary of the local, told the assembly that with their permission, a district officer could call a strike. The assembly so voted, and then gave the necessary authority to the only qualified man on the scene, James Kirwan.[27] The only explanation given for this unusual procedure is that the union doubted its ability to obtain the necessary three-fourths majority from a general referendum of all members.

On the following day, before Kirwan could call the strike, one of Grier's terse notices appeared in the newspapers: "The Homestake Mining Company will cease operating its property this evening."[28] The shutdown began at 6 P.M. For the first time in thirty-two years, the Homestake was not in operation. The immediate sensation was the overpowering quiet as 800 stamps ceased their clatter. "You could even hear the dogs bark," recalled one awed observer.[29] The people were fearful and tense but did not panic as their futures suddenly became insecure. Ironically, the next day was Thanksgiving.

The decision to use the lockout was obviously not a hasty one. For two weeks preceding the order the company had been preparing the mines and mills for the shutdown. The men moved the powder supply to a higher level of the mine, barricaded the tunnels, stopped the pumps, and made a general cleanup in the mills.[30] Chambers Kellar directed the Homestake campaign against the union from this point on. His office on Mill Street was the command post during the entire lockout.[31] Grier was still the ultimate authority, and Kellar worked closely with him.

As soon as the lockout was declared, miners prepared to move from Lead to other camps where they hoped to find work. Their choice was limited. Butte, Montana, as usual, was preparing for labor trouble, and many of the mines in Canada and Alaska were in the throes of strikes. A number of the miners went to Globe, Arizona;[32] in fact, within a week the mayor of Globe wired that no more should come as his city had too many unemployed.[33] The same situation was true in most of the western mining camps, and similar notices appeared for months in Lead newspapers. The first miners to leave were the footloose drifters prevalent in gold camps; the established married men would be going in a short while. The social structure of Lead was starting to distintegrate.

While the first miners were leaving, a new group of men were arriving. The Homestake imported Pinkerton, Thiele, and Boyd detectives from Denver. The Hearst Mercantile armed them with riot guns and pistols, and the sheriff of Lawrence County swore them in as special deputies.[34] Union supporters referred to them as "thugs and murderers"[35] and as men "with notches on their guns."[36] The *Register* felt that they were coming to the camp to create an incident that would allow the Homestake to ruthlessly crush the union by sheer force.[37] There is little doubt that the quality of the men employed by the detective agency was low. Many of them were vagrants and other undesirables hastily recruited in Denver; some were quite ordinary people looking for jobs.[38] The company used them only to protect its property and to guard its officers and their families.[39] They were never used to attack union members by force, if for no other reason than that such action was not necessary. The city of Lead, however, presented the appearance of an armed camp. At night a large spotlight, placed on the top of the Ellison hoist, played over the city and the company works as it sought to pick out the congregated groups of men who might cause trouble.[40]

Three days after the lockout the W.F.M., through the actions of national Vice-President C. E. Mahoney, succeeded in alienating a large number of its men. This grew out of an incident in a local cafe. Mahoney overhead Father Chasse, a Catholic priest, make the statement that he felt there would have been no lockout except for the influence of the national W.F.M. Mahoney verbally attacked Father Chasse in language that was explicit and profane. Someone informed Mahoney that he was talking to a priest, but Mahoney replied that he did not "care a damn for any damn dirty priest."[41] The incident received considerable coverage in the newspapers, and the next night the Knights of Columbus held a special meeting that condemned Mahoney and demanded that the union withdraw him. Mahoney was arrested and fined $50. He left the next day, but the damage had been done. At the same time the local union was having internal trouble. The president of the local, a Mr. Rowe, was violently opposed to the strike, which by now had been called by Kirwan. He left Lead as did Mr. Bant, a trustee of the local. Neither returned. The split in the union ranks was becoming quite obvious.

Charles Moyer arrived in Lead December 2, 1909, to inspect the situation. The next night the local asked Moyer to provide benefits for the strikers. Moyer told the men that they must use their own resources to exhaustion before asking other workers to help support them. He said that a special assessment of $1 per member was being made for their benefit but that the funds would not be available until January. He advised the men to make a "peaceful fight."[42] At this time union officers were looking for places that would board men cheaply.

Moyer may have regretted his trip to the Black Hills. The next day the *Lead Daily Call,* privately owned but with a strong company bias, ran a story that said Moyer had been caught the previous night in a raid on a local bordello. It further stated that upon his release, he went to yet another house of ill fame.[43] The story was undoubtedly false, but it hurt Moyer's reputation and the local union effort.

On December 9 the union called out 300 watchmen and maintenance men not affected by the original shutdown of the mine. Detectives, who had been doing the job anyway, filled their places. At a mass meeting on December 13 there was talk of attempting to arbitrate the strike, but nothing came of it. This meeting called out the drivers who worked for the Hearst

Mercantile Company. The union treated the company and the community to a show of strength on December 15. One thousand men paraded down the main street to the accompaniment of scattered catcalls and boos. The march was quiet with no untoward incidents. It failed in its purpose as a show of strength. Company supporters noted many men from other camps among the marchers and also noted that 1,000 marchers out of 2,500 employees was a low percentage.[44]

Toward the end of December the Homestake began publishing in the area newspapers copies of cards that any future employees would be expected to sign. These cards read as follows:

HOMESTAKE MINING COMPANY:
Lead, South Dakota .... 19......
I am not a member of any labor union and in consideration of my being employed by the HOMESTAKE MINING COMPANY agree that I will not become such while in its service.
....
Department ....
Occupation ....[45]

This was a revival of the card system of Cripple Creek. The *Miners' Magazine* compared the Homestake to the Czar of Russia for adopting it and stated that the company had no generous feelings toward its employees.[46] *The Lantern* said to the men, "Dare you sign that card and look your wife and babies in the face?"[47] The union had reason to fear the card system. The adoption of this system was a clear indication of the Homestake's determination to break the union completely.

The union attempt to break the lockout depended to a large degree on whether it could keep the loyalty of its members. In trying to do this, the W.F.M. concentrated its greatest effort on the large number of foreign-born workers. Generally the native-born Americans would desert the union before the foreign-born. Of the latter, the largest group was the Slavonians, big raw-boned men from Yugoslavia. There were at least 700 of them in the union. The efforts of Terzich, the Slavonian-speaking W.F.M. leader, held them.[48] Terzich not only had a natural empathy with these men but also organized very effectively. If it appeared that one Slavonian was wavering, the others would write to his relatives in the homeland urging them to write the dissenter to hold fast.[49] The majority of Finlanders also tended to hold fast. The Finns more than any other group included dedicated socialists. Not all Finns supported the union of course, but the percentage was high. Later in the strike the Finnish Temperance Union expelled a member for working for the company.[50] Robert Bertolero was the leader of the Italians, who showed less loyalty to the W.F.M. than the Slavonians or Finlanders. The Cousin Jacks from Cornwall, England, assimilated very easily into American society, and efforts to approach them on a group basis were difficult. As the most professional of all the miners, they tended to have the better jobs; hence it was extremely unlikely that they would work up any great enthusiasm for the union. The acid test of loyalty did not come in the first weeks of the lockout but later when the savings were gone. It was then that the abstraction of union loyalty had to face the reality of economic necessity.

It was essential that the union keep its workers in Lead in order to avoid total surrender to the Homestake. Beginning in December 1909, money poured into Lead from various union sources at the rate of $6,000 per week. A children's Christmas donation brought in some money. Individuals and unions all over the country sent in such funds as they could.[51] The Butte Miners' Union No. 1 sent a check for $25,000.[52] The largest amount came from the assessment funds collected by the national W.F.M. Strike relief amounted to $202,430 by the 1910 summer convention of the W.F.M.[53] The convention of 1911 reported that $228,832.25 had been spent during the preceding year.[54] It was not until 1912 that the W.F.M. ceased the contributions. These large sums of money, paid by a relatively poor union, show the importance that the W.F.M. attached to the struggle in Lead. No other union received anything like this support during this period.

While the basic struggle in the lockout was between the Homestake and the W.F.M., other groups and individuals became participants by choice or necessity. The newspapers took sides early and fought to the end. The union claimed that the press was in the pay of the Homestake.[55] It is true that the *Lead Daily Call* was extremely procompany and earnestly reflected the Homestake views on all occasions. The editor, W. R. Grace, and his wife, Belle, who wrote the majority of the editorials, were conservative Republicans; the radicals and socialists in the W.F.M. ranks horrified them.[56] Early in the strike Grace hired an armed guard to protect him but never ceased to pour invective on the union. The union also had considerable press support. The *Black Hills Daily Register*, edited by W. C. Benfer, a former W.F.M. official,[57] called itself a radical newspaper and supported both unionism and socialism completely. The most colorful paper supporting the union was *The Lantern*, published in Deadwood. Freeman Knowles, the editor and publisher, cared little for either propriety or the law of the land, although he was also a lawyer. He went to jail in August 1909 for failing to pay a $500 fine occasioned by a breach of the postal laws: he had published his views, rather advanced for a seventy-year-old Civil War veteran, on sex and marital relations.[58] Released when friends paid the fine, he entered into the lockout with such relish that by February 1910, he was appealing for funds with which to defend himself against three libel suits.[59] He was a superb gadfly and worked with all his fervor for the union cause. The combination of high excitement, work, and jail proved too much for the old man, however, and he died June 1, 1910. The union that he had so well supported eulogized him, as was his due. The remaining newspapers in Deadwood and the other Black Hills communities were largely procompany, although their attitudes were more detached and their comments less personal than the Lead papers.

The roles of various levels of government during labor troubles are difficult. This was the case in Lead. The elections of 1908 had been a narrow victory for the Republicans, largely supported by the Homestake.[60] This election, a year before the lockout, may have been the most significant political act that affected the lockout. Had the union elected a ticket headed by Freeman Knowles, candidate for State's attorney, the situation during the lockout could have been considerably different. The union claimed that local, county, state, and national elective officials were persecuting them.[61] None of the evidence indicated this, however; in any case, persecution was unnecessary. Neither side wanted violence, nor did the agencies of government.

There were definite problems for the city government.

An attempt by Alderman Scoggin to raise the city funds available to the needy was voted down because the majority of the council felt this might be interpreted as strike relief. Later the Socialist party in Lead expelled Alderman E. D. Thomas for going over to the Homestake. Thomas resigned from the city council because of this, although the other council members urged him not to.[62] There were debates on the propriety of allowing the company to bring its special deputies into Lead. The council reached no decision, and the deputies remained.

The union made a last attempt to gain political control of Lead in the April election of 1910. A Union-Labor ticket ran against a Citizen's ticket backed by those supporting the Homestake.[63] The Citizen's ticket, headed by Harry L. Howard, candidate for mayor, won by a majority of 269 votes out of 2,200 votes cast. The union, which ran a good race in the traditionally conservative Lead, elected the third-ward alderman,[64] due to the large Slavonian and Finnish population in that ward. The union charged fraud in the election, stating that the Homestake rigged the polls and imported voters.[65] They charged that the local and county officials continued to persecute them.[66] All attempts by the union to find a political solution had failed.

While the W.F.M. was the prime target in the lockout, the Homestake's nonunion position affected other unions in Lead and surrounding camps. Grier made it clear that the order applied to all unions.[67] Because the W.F.M. included most of the industrial workers, the order affected very few other people, but other unions made common cause with the locked-out miners. Barbers, bartenders, musicians, teamsters, carpenters, electrical workers, and working girls all contributed to the W.F.M. strike fund.[68] They felt that the destruction of the W.F.M. would mean the end of all unionism in the Black Hills. Strangely enough, the *Call* made no attempt to bar A.F. of L. printers from its plant. Editor Grace justified this on the grounds that the A.F. of L. was not socialistic in nature and that no one would object to a union if the socialistic elements were removed.[69] The various socialist locals in Lead, Central City, and Deadwood were enthusiastic supporters of the W.F.M., although their value may have been exceeded by the wrath they drew upon themselves.

One of the few things on which the Homestake and the

W.F.M. agreed was the Industrial Workers of the World. Both despised the I.W.W. In September 1909 John Walsh, an I.W.W. organizer, came to Lead to organize the camp. The *Call* attacked him and his followers as "shouting blatherskites" and suggested their arrest.[70] The W.F.M. was more emphatic. James Kirwan confronted Walsh and ordered him to leave town before he was injured. Walsh did. The W.F.M. regarded the I.W.W., its former partner, as having "degenerated into a howling mob lead by a few irresponsible freaks."[71] The W.F.M. refused to recognize the I.W.W. as a bona fide union and would not accept or recognize I.W.W. union cards.[72] The situation in Lead proved that dual unionism between the W.F.M. and the I.W.W. was finished.

The businesses of Lead, Deadwood, and nearby communities felt the impact of the lockout very early. As soon as the lockout was declared, retail firms launched going-out-of-business sales, preparatory to leaving the Hills. The remaining businesses found themselves inexorably forced to take sides in the contest. The union charged that the local tradesmen, together with the doctors and the lawyers, were trying to subvert union members.[73] To counteract this, a group of fifty-three firms (mainly small, family businesses run by Slavonians and Finlanders) banded together to condemn the Homestake's actions and to support their customers, the union members.[74]

The Homestake deliberately courted business support and received most of it. The other mining companies operating in the Black Hills concluded that if the Homestake, which employed three-fourths of the men and produced 90 percent of the gold in the area, was going to operate with nonunion labor, they had better do the same.[75] According to Grier, the management of these mines approached him and offered to join in common cause against the unions.[76] He accepted the offer, and these mines shut down in January 1910. They announced they would not reopen until they could do so with nonunion labor.[77] The Homestake and the thirteen other companies involved ran a continuing advertisement that read as follows:

> In view of the fact that the Mining Industry in the Black Hills District is the source from which all other business interests in said District derive their main support, and that said industry intends to establish, permanently, in said District what

are commonly called Non-Union labor conditions, it is respectfully suggested to all such other business interests that their actions should be vigorously in support of the aforesaid expressed intention.[78]

The *Call* regarded this coalition as the doom of the Western Federation of Miners in the Black Hills.[79] It certainly brought the full force of the mining industry and a considerable portion of the business interests of the Hills into play against the union.

With the high passions aroused during the lockout, the small amount of violence connected with the strike was surprising. This was in part due to the changed outlook of the W.F.M. and also to the efficient law enforcement in the region. Law enforcement officials expanded their departments and handled the few incidents that arose much as they would any other law violations. No special methods, such as the bullpen and the deportation, were adopted. There was apparently no thought of asking the governor for the militia or the president for soldiers. Local law enforcement handled the situation, while the Pinkertons confined themselves to guarding company property and bodyguard work.

There was of course some violence connected with the lockout. The union accused the Pinkertons of beating five union members but could not prove the accusation in court.[80] There was little doubt that occasional fights occurred in the saloons of the town. An order by the sheriff and the state's attorney prevented the carrying of concealed weapons, although the union protested that it was discriminatory.[81] This order doubtlessly prevented much serious trouble. One Alonzo Goss allegedly had red pepper thrown into his eyes by an unknown assailant who said, "Get out of my way, you scab."[82] Unknown persons left several bottles of phosphorus with fuses in them in Lead, and one phosphorus bomb exploded behind the Homestake Hospital.[83] Fortunately, no one was injured. In the one attempt to destroy Homestake property, Jack Butler, a loyal union man, threw an incendiary bomb from a moving train into the cyanide mill at Whitetail. The bomb started a fire which was put out before any damage was done.[84] Officials arrested Butler, who was tried and sent to the penitentiary.[85] Apparently Butler did this of his own volition without orders from the union. The most spectacular bit of violence involved Chambers Kellar and Freeman Knowles. Kellar, apparently quite incensed about some of the comments appearing in *The Lantern* about him, waited in the county commissioners' room of the Deadwood Court House for Knowles. When the old man entered, Kellar pulled out a riding quirt and struck him several times. In the ensuing struggle, Knowles blacked Kellar's right eye.[86] The state's attorney watched the fight with bemused interest but did nothing to stop it and made no arrest. Knowles never pressed charges against Kellar. He contented himself with several biting articles about the "chief pugilist of the Homestake Mining Company."[87] The union was on its best behavior during the lockout, and William Tracy was quite correct when he said, "The union could have destroyed the Homestake works, but didn't."[88]

In January 1910 the final moves commenced, resulting in the destruction of the W.F.M. in Lead. On January 6, forty-nine men met in Society Hall and formed the Loyal Legion. George D. McClellan was the president; W. J. McMakin, vice-president; Will Treweek, secretary; and William Royce, treasurer.[89] Freeman Knowles derided them as "cripples, deadbeats, pimps, preachers, doctors, lawyers, and school children."[90] The organization also included miners and soon had more of them. On January 1 a committee from the Loyal Legion, wearing appropriate buttons and ribbons, asked Grier for permission to go back to work. Grier, after thinking over the proposition for a day, agreed and announced on January 10 that the mines and mills would reopen.[91] The Loyal Legion grew rapidly and attracted many men from the union ranks, particularly the more highly paid men.

As soon as the decision had been made to reopen the mine, the Homestake started to look for laborers. Agents were sent to mining regions all over the United States to recruit scab labor. By January 15 the first scabs arrived.[92] The union claimed that Grier had promised not to import scabs (although why he should so promise was not explained) and considered taking legal action to enforce the alleged agreement. The greatest number of scabs came from the lead-mining regions of Kansas and Missouri. The depressed condition of the lead-mining industry had made them susceptible to the blandishments of the Homestake's agents.[93] Scabs also came from Colorado, Michigan's Upper Peninsula, North Carolina, Wisconsin, Georgia, and Tennessee. The company paid their fares, found them places to live, and extended them credit until they were established.

Some of the company agents on recruiting duty had troubles. Bruce C. Yates, later to be superintendent, was hiring scabs in North Carolina and Tennessee, unaware that a Tennessee law forbade such action. When he heard that a warrant had been issued for his arrest, he hired a rowboat and crossed the Tennessee River into Georgia at 1:00 A.M.[94]

The union claimed that the scabs were incompetent.[95] It produced articles written by dissatisfied scabs and sent them to the regions from whence the scabs had come.[96] These tactics availed them little, however; the scabs still came. Landlords in Lead squeezed out union men in order to move in the outsiders, whose prospects for having ready money were infinitely superior.[97] Lead was splitting into two factions. The scabs or nonunion workers were in active competition with the rednecks or union men. The ill feeling between the two groups extended to the children as well as the adults and lasted many years.[98]

The same day that Grier announced the reopening of the mine and mills, the Homestake made the first preliminary moves toward the actual reopening. A force of machinists started to ready the equipment, and the mine horses returned to town from the pasture. Company locomotives shifted supplies. Five hundred men had already joined the Loyal Legion,[99] and on January 13, 1910, the Star hoist was operating and lowering the first men into the mine to work the pumps. When these men came up at the end of their shift, the streets of Lead were packed with curious citizens and nervous police. The men were somewhat afraid to leave the company grounds, but when they did, there was no trouble.[100] By January 19 the big Ellison hoist was raising ore mined before the lockout; two days later the Amicus stamp mill started its 240 stamps pounding. By the first of February, 640 stamps at the Amicus, Star, and Highland mills were at work. The company was hoisting at the Star, Ellison, and Old Abe hoists and mining in the open cut.[101]

Each day more men deserted the union and went to the Homestake offices where they surrendered their union cards and signed the statement guaranteeing that they would not become union members while in the company's employ. *The Lantern* of February 10 estimated that 1,000 men had signed while 1,200 still remained with the union. Practically all the men still on strike were Slavonians, Italians, or Finns.[102] However, William Benfer estimated in March that fewer than 200 had deserted the union although another 500 had left the camp. He thought that 1,600 men in Lead, mostly miners and muckers, still stood firm for the union.[103]

By the middle of February the Golden Reward and Mogul mines, allied with the Homestake against the union, reopened their Bald Mountain operations with scab labor imported from Michigan.[104] At the end of February all of the smaller companies were approaching full production. The last of the Homestake stamp mills, the Deadwood-Terra, started operation on March 3, putting the company into full production.[105] The lockout, for all practical purposes, was over. Grier estimated that over 1,000 former union members were working for the Homestake.[106]

The Western Federation, although faced with the sight and sound of the company at work, refused to believe that it had lost. The union boasted that it could feed its men forever and that the Homestake was suffering enormous losses maintaining a sham of full production.[107] It claimed that the miners had not deserted the union and that several hundred scabs were working the plant alone.[108] This attitude not only was unrealistic but also hurt many men who, believing that the union would still win, waited too long. When they decided to return to the Homestake, many found their jobs already filled and had to move on.[109]

Some of the union leaders deserted the camp. Chris Christianson, "the blacksmith orator" of Lead, left for the Wyoming coal fields in March. In December 1909 he had said, "No, we won't settle. . . . The Homestake has run us a long time; now we are going to run the Homestake; boys it belongs to us, and by God, some of these days we'll take it."[110] Tracy and Terzich left in April. The local leaders, such as Kirwan and Ryan, stayed and continued to fight.

The W.F.M. persisted in the struggle. It continued to pour relief funds into Lead through 1910 and 1911. The executive board, visiting Lead in December 1910, could still see hope.[111] The committee on strikes urged the 1911 convention to continue the struggle in the Black Hills.[112] The delegates from the Black Hills, however, felt that they could not hold on much longer. By 1912 the W.F.M. was still warning workers to stay away from the Black Hills, although the strike relief had stopped. Lead City Union No. 2 no longer had a place on the list of W.F.M. unions.[113] The local was out of money and only a few men remained of the previous 2,000. By 1913 the W.F.M. had canceled the charters of all Black Hills locals except the small local at Galena.[114] That same year the union paid transportation charges for the last of the men leaving Lead, as Charles Moyer admitted that the Homestake had won.[115] By 1914 all the unions in the Black Hills, both A.F. of L. and W.F.M., were gone.[116] The Homestake's victory was complete.

The Lead City Local dragged on to a shabby end. Moyer had removed its charter on January 3, 1913, for nonpayment of dues. Tom Ryan and fourteen others stayed on and controlled the union hall until the following December when the national organization took over and locked Ryan out. A lawsuit followed.[117] Eventually the hall became the property of a bank. Ryan was understandably bitter toward Moyer and joined in the fight against his leadership.[118]

The merchants who had come to the support of the W.F.M. publicly were in an embarrassed position with the end of the lockout resulting in a Homestake victory. The company itself had never moved against them, although it easily could have put any or all of them out of business by revoking their licenses to occupy Homestake land. The *Lead Daily Call,* however, published its lists of heretical businesses and argued:

> It is and always has been the inalienable right of the Homestake employee to do his buying wherever he sees fit, but with so many reliable business men in the city who are in full sympathy with the Homestake and its workingmen, isn't it a mistake to give encouragement and support to your enemies?[119]

The editor's syntax was unable to keep pace with his righteous indignation, but his point came through clearly. In 1910 the heretics saw the error of their ways and went through a sort of absolution by entering notices such as this in the *Call:*

Editor of *Call:*
Dear Sir:
Will you please take my name from the list headed
"friends of the Union" now running in your paper? I
erred in taking sides in the first place, and I am sure I
have nothing against the Homestake company or the men
now working.

K. ZOULIS[120]

This apparently took care of their sins. Many of them—John
Mayo, William Bertolero, and John Mastrovich for example—
were prosperous and respected citizens forty years later. Others
who were unable to withstand the financial loss when their
customers left disappeared from the community.

The *Engineering and Mining Journal,* in an article com-
paring Butte (a completely unionized town) with Lead (a nonunion town), found that Lead had "incomparably superior
physical conditions . . . the only condition is that the miner
shall have nothing to say about it himself."[121] Professor John R.
Commons, chairman of the Commission on Industrial Relations which investigated the lockout in 1914, complimented
Grier on his policies and called the Homestake "benevolent
despotism."[122] Even a bitter enemy of Grier and the Homestake, Bishop Joseph F. Busch, stated that "in Lead we had a
situation that was from a material standpoint rather preferable
than most."[123] Nevertheless, Bishop Busch deplored "the all-
pervading influence of the company" and felt that there was
no independent sentiment in the community.[124]

The Western Federation of Miners had found no answers
in Lead. It had attempted to forego violence and to use hon-
orable bargaining methods in its place. These had failed for
several reasons, the most elementary being the tactical incom-
petence that allowed the W.F.M. leaders to be drawn into the
lockout without any immediate cause for which to fight or any
means with which to fight. The Homestake outmaneuvered
the union at all turns and did not have to use many of the
means available to it, such as the removal of strikers' homes
from company grounds, the injunction, and the calling of
either state militia or the U.S. Army.

Of more importance than the tactical reasons for the defeat was the general state of labor-management relations at
that time. In the period from the Civil War to the 1930s unions

were seldom successful in their disputes with management. Public opinion was usually against them. Law enforcement agencies on all levels of government were more concerned with protecting company property rights than in any set of union principles. The courts, both state and federal, interpreted the statutes in such a manner as to make the injunction a usual device of management and to prevent the unions from having any real recourse at law. There was no provision for compulsory arbitration, no recognition of a right to negotiate, and no right to either a closed or a union shop. Under such conditions any union was comparatively helpless against a strong, determined adversary.

The Western Federation of Miners had succeeded on occasion by the use of violence. The sheer terror it inspired could sometimes overturn the odds against unionism. In the long run, however, violence worked against the union by bringing the full wrath of the public and its government to bear on all unions. Violence and radicalism were not the answer. However, when Moyer withdrew from violence he had no weapon to replace it. In Lead the W.F.M. could only fight with words, and that was not enough to counter the Homestake's very real power and its record of fair dealing.

The Homestake Mining Company emerged stronger than ever from the lockout. There was no force capable of challenging its power over the workers and the community. This power probably carried with it temptations to turn toward a more authoritarian regime regarding labor. If so, the company resisted and instead increased its welfare activities and assumed responsibilities toward the worker that it had not previously held.

CHAPTER EIGHT

CULMINATION OF THE SYSTEM: 1910-42

THE HOMESTAKE had effectively crushed the Western Federation of Miners and unionism in general in the northern Black Hills mining camps. The experience culminating in the lockout convinced the company that unionism was an evil to be avoided at all costs. The men who ran the company, as well as the vast majority of the citizens of Lead, would forever identify all types of unions as anarchistic and destructive. This feeling would pervade all thinking and all actions regarding the treatment of labor ever after.[1]

With a strong bias against unions, it might have been understandable for the company to turn to strong repressive control of its working force. Selfish, antisocial individualism and commercialism were common to the period and to the industry. Yet there is not the slightest evidence that the company considered such policies. The harrowing experiences of the lockout failed to shake the management's faith in its basic methods. Its paternal social conscience would not allow a change from the system that the company believed would create a peaceful, class-harmonizing situation. Basically conservative, it would build on that which was traditional. The Homestake was willing, even eager, to expand the services to the employees. The removal of all charges for hospital service during the lockout illustrates this well,[2] as does the policy of keeping the hospital open and operating for the locked-out employees during a period when they were not working.[3] Indeed, the destruction of the union required that the company take over some of the necessary functions that the union performed. Consequently, the company programs were somewhat easier to install, and acceptance was more or less insured by the exile—voluntary or enforced—of those totally bound to the union. There was no hard core of opposition left in the community.

The philosophy of the Homestake's welfare programs and its general policies toward its employees were best stated by Superintendent Bruce C. Yates in 1920:

> Men and women, to be happy and contented, should have definite work to do, have amusement, and must be allowed to develop their home and community life in their own way. Welfare work, to accomplish the most good, must help the people of a community in their daily work, their amusement, and their homelife. It should be developed gradually and naturally as the needs of the community are manifested and should never be conducted along fixed lines laid down by a single individual who perhaps lacks full appreciation of human nature. Paternalism should be avoided as far as it is possible to do so in a community where a single business interest dominates. The needs of the people should be carefully studied and anticipated; and the one central idea that each individual employee is part of the organization—that he is not a machine but a human being with the usual characteristics of the species—must govern if the work is to be successful.[4]

Yates's comment on the avoidance of paternalism was apparently meant in a special sense. Certainly the entire policy of the company toward its workers was paternalistic. Yates probably meant that the policies should not be forced on the workers, that the workers should retain the freedom of choice in them, and that the welfare should be built on manifest need and community traditions rather than a priori theories.[5] Yates was not developing any new theory of welfare work himself but was merely generalizing on what had been done previously and doubtlessly would continue.

Following the lockout the company moved with all deliberate speed to fill the gaps caused by the demise of the union. The union hall had been the community center; now it was closed and in litigation.[6] The company apparently started its planning for a replacement shortly after the lockout, and by 1912 workers were constructing the new recreation building.[7] The building, a three-story brick and stone structure costing $250,000, opened for public use on August 31, 1914.[8] The first floor of the building contained a lobby, a card room, and billiard tables. The basement held a gymnasium and bowling alleys. The upper floor was the home of the library. Attached to the building was a theater with a seating capacity of 1,000; both motion pictures and legitimate theater and vaudeville played there.[9] The admission prices for the theater were deliberately kept low. During the inflationary period in 1919, children could attend for five cents while adults paid fifteen cents.[10] Beneath the theater was a 25 × 75-foot tiled swimming pool. The pool area included change rooms, showers, and even hair-drying facilities for the patrons. The water was heated, filtered, and disinfected.[11] There were lifeguards and various other attendants.

With the exception of the theater, all the facilities of the recreation building were absoutely free. No membership fee was charged and no card system used. Anyone in Lead could make full use of it,[12] company employee or not.

The company's recreation program was not limited to the activities performed in the building. The Homestake Recreation Department helped the community in other ways. The local Rod and Gun Club received a clubhouse for its exclusive use. The original cyanide mill became a skating rink after dismantling. Rifle and pistol clubs were allowed the use of company buildings that were no longer needed. The department operated basketball and baseball leagues. When agitation arose for a country club in the 1920s, the company leased land to the club which had originally been acquired for water rights, loaned it money, and had Homestake crews provide specialized labor for construction and repair. In return, the club kept prices for membership low and membership available to all who were interested.[13] The company provided similar help to a ski club in the 1930s. It also purchased a large club building on a beautiful site in Spearfish Canyon for the free use of any organization notifying the department of a desire to use it.

The company always allowed the public schools full use of its facilities. The schools taught physical education classes in swimming in the Homestake pool. In 1919 the Homestake provided the schools with an athletic field by blasting a level area on a mountaintop.[14]

The company's record in recreation is most impressive when one considers that the field of industrial recreation did not develop nationally until the late 1930s and that most such programs were and still are on a fee basis.[15] The methods used were equally impressive. In accordance with Yates's philosophy, no one was forced to participate. The company tried to assist the employees in their activities rather than to establish the activities and force the employees into them. The citizens of Lead ran the bowling leagues and the golf course. After the demand for an activity gained impetus, the company would provide help. In the case of the basketball leagues, the players organized them and the company then provided officials, equipment, and a place to play.[16] Changes in the programs occurred as old activities lost their appeal and new ones gained support. The company, after operating its theater for years, leased it to the Black Hills Amusement Company in the 1930s with the provision that the prices be kept low.[17] The general philosophy in the recreation area saw no change. The recreation building employed men who were no longer able to work in the mines and mills. It also provided employment for boys as pinsetters, lifeguards, and part-time assistants. The department avoided buying automatic pinsetters in order to keep this employment for the boys, although it was costly to do so. Recreation programs helped to keep the workers satisfied and also achieved a degree of social cohesiveness in Lead. There was a focal point for the social and cultural activity of the town—a common meeting ground for manager, worker, and townsman that enabled all to know and to understand each other.

In addition to the general area of community culture, the union had provided benefits for ill and disabled workers. The Homestake moved into this area rapidly by establishing the Homestake Aid Fund Association August 1, 1910.[18] A five-man board of directors, elected annually, operated this fund. The Homestake superintendent served as treasurer and the chief clerk as secretary. Each Homestake employee contributed $1 per month to the fund while the company contributed $1,000 per month. The fund originally paid benefits for both sickness and accidents, but the passage of the South Dakota Workmen's Compensation Law of 1917 required the company to pay all the costs of accidents; after that the fund was used exclusively for sickness.[19] Originally, $1 per day was paid for loss of time, beginning immediately after an accident and after the sixth day in case of sickness and continuing for a maximum of six months.[20] Later the benefits increased to $1.50 and the time to nine months. The association paid $800 to the beneficiaries in the case of death and $200 if the member committed suicide or became insane.[21] Great care was taken to insure that the fund was well administered and that the employees should understand this. Monthly reports of benefits paid were published in both Lead and Deadwood newspapers.[22]

The company stood behind the fund at all times. The influenza epidemic in 1918 put great pressure on the system. In the month of November alone, twenty-six died, and the fund paid $20,800 in death benefits.[23] The fund had a deficit of $7,000 that the Homestake assumed.[24] In addition to the fund, the company also liberally supported two charitable organizations established by its employees. The Ladies' Auxiliary to the Aid Association, which had a membership of 150 in 1920, helped families of employees unable to help themselves.[25] The Homestake Veterans' Association, formed in 1905 and composed of 21-year men, conducted charitable activities among its members and other workers.

In 1917 the Homestake began the practice of retiring its long-term employees because of old age or physical disability and started a pension system. The employee was paid 25 percent of the last full year's pay plus $10 per year for each year's service with the company. The total could not exceed $600 per year.[26] The payment was nearly always $50 per month; few received less.[27] By 1920 forty-two men were on pension, and the number increased thereafter. Pensions were also paid to widows of employees killed in accidents.[28] The pension system was continued even after the passage of the Social Security Act in the 1930s and was used essentially to supplement the benefits of the act.

In addition to these formal benefits, informal arrangements were made to care for the workers. A man who was unable to stand underground work would be transferred to the surface, or perhaps a surface worker would be shifted to easier work. The employment records of the company's workers in 1936 show that a man with 20 years of service would have had an average of 2.6 transfers; nearly all of these were from hard jobs to easier ones or from the dangerous to the safe.[29] Surface employees transferred less often than those underground.

At 4:30 P.M. on April 13, 1919, Phoebe A. Hearst died in her home in Pleasanton, California, a victim of the influenza epidemic still sweeping the country.[30] She had been a firm friend to the workers and to the community of Lead. Her death did not mean the end of her work, however, because the company assumed the support of all her personal projects in the community. She had done her work well, and company policy would forever be in her image.

Among Phoebe's many activities was that of "Americanizing" the immigrant people who worked for the Homestake. This had been a prime objective of the kindergarten which the company accepted and continued. The company also regarded the recreation building as fostering Americanization.[31] In 1919 the program was stepped up, possibly as a result of the hyperpatriotism engendered by World War I and also probably as a result of a desire for greater efficiency that could be obtained by exclusive use of the English language. In an open letter to the employees dated April 1, 1919, Yates told the workers that the Homestake was an American corporation and that it wanted all its employees to speak and write the English language. To enforce this, all bulletins and notices would henceforth be written in English only. The company intended to discourage all conversation in a language other than English and urged the employees to have their families learn English as soon as possible, stating that the "Homestake Company does not wish to retain in its employ citizens of foreign countries."[32]

The company kept records of the foreign-born and closely kept track of their road to citizenship.[33] The local school district ran night schools for the foreigners where English and other subjects necessary to pass the naturalization tests were the curriculum. The courses were free and had "an attendance that is growing every session."[34] The employment department sent lists of men who intended to take out citizenship papers to the schools, and the two worked closely on the program.[35] It was extremely effective and perhaps necessary in a community like Lead. Certainly the company regarded this as welfare work and as being in the best interests of the workers.[36] Yet it can also be regarded as the sort of paternalism that Yates tried to avoid and an intolerable invasion of private rights. The Homestake Mining Company did not regard it so, and there is no evidence that the workers objected.

The Homestake continued and expanded the services that it had been operating before the lockout. This was particularly true of the medical department. In addition to removing the charges for medical and hospitalization services in 1910, the company expanded them to include general medical, surgical, and obstetrical care to employees and their relatives.[37] Approximately 8,000 people were eligible for the free care. In 1920 the hospital employed seven full-time physicians in Lead.[38] In addition, a physician was stationed at the timber center in Nemo and later at Spearfish, while a part-time doctor was employed to care for company employees at the Homestake's Wyodak Coal Company located in Gillette, Wyoming. The company hospital was replaced in 1923 by a modern brick building.[39] The old hospital was never actually destroyed; the new one was built around it and the old building is still the core. The company insisted on high standards of patient care, and only graduate registered nurses were employed.[40]

The hospital department handled an ever-increasing patient load. In 1924, for example, 12,960 house calls were made, and 43,500 dispensary patients and 2,223 surgical cases were treated.[41] By 1929 there were 12,187 house calls, a slight decrease, but the dispensary load had risen to 48,359 cases.[42] By 1940 the house calls had risen to 16,150 while the dispensary patients jumped to 89,625.[43] Only pregnancy remained stable during this period. There were never fewer than 121 nor more than 145 confinements. The complexity of the medical practice increased somewhat faster than the patients.

One of the problems facing the medical men concerned with hard-rock miners was silicosis. In 1936 they examined 1,915 men in an attempt to pinpoint this disease; 201 cases of silicosis (10.4%) were found. Of these men, 127 (63.18%) had worked in other mines. In the mining department itself, 161 of 1,140 workers examined had silicosis. The chief surgeon, Dr. R. B. Fleeger, concluded that the disease existed only in the mining areas and that the methods of controlling it were inadequate. He recommended a reduction in heavy blasting, sampling of the dust content of the air, use of respirators when possible, and the removal of diseased employees to surface jobs.[44] Nothing seemed to help, however. By 1965 there were 160 workers who had the disease. One of the probems was diagnosis. Silicosis cannot be determined until it is in a moderately advanced state.[45] Nothing seemed to prevent it. The company provided respirators, water drills, and periodic chest X rays which were required. These preventives seemed to help little more than the oldtime preventive of the miners—

the wearing of a luxuriant mustache to filter the dust.[46] The disease struck indiscriminately. Two miners could work side by side on the ore face, and only one would be affected. One might acquire the disease in four years, another might take twenty. The average was sixteen years' exposure.[47] It seemed that the "miner's con" was to be the fate of those who drilled hard rock. The Homestake could not prevent it, but neither could anyone else.[48]

By 1942 the Homestake medical staff and the hospital were an efficient and smooth-running operational unit which provided excellent service to the workers. Gaps existed in the program, however. An employee or his dependent who was injured elsewhere had to take care of the expense himself if unable to get to Lead.[49] On the other hand, the employee frequently took advantage of the company by bringing in sick relatives from long distances for the company's hospitalization services. On balance, the Homestake employee had superior medical care.

The Homestake had always been a relatively safe mine. It was well engineered and scientifically worked. Still there were accidents. The company rhetorically stressed safety and no doubt desired increased safety for its workers at all times. In 1917 it did something about it. Bruce Yates stated: "In keeping with the policy of the company of looking after the welfare of its employees, a safety engineer was added to the staff in 1917, and a regular organization effected to look after matters of safety."[50] There can be no doubt that the Homestake was concerned with the welfare of its workers, but to some it may appear to be more than coincidence that 1917 was the year of the passage of the South Dakota Workman's Compensation Law that required companies to pay for injuries occurring on the job. Safety was welfare, charity, and philanthropy prior to 1917; after that date it was business and more efficiently handled. Regardless of motivation, safety did benefit the worker as well as the company.

The company's approach to safety became systematic and effective. The company joined the metals division of the National Safety Council in 1917 as one of its eighteen members.[51] A workingman's committee, chaired by the safety engineer, met once a month to discuss the specifics of safety in the mine. If something needed immediate attention, the engineer could act without committee approval.[52] To encourage and promote safety, the company published *The Homestake Safety Bulletin* on a bimonthly basis. It was filled with practical tips and gory warnings of the fate awaiting the careless worker. The illustrations were enough to drive a man to another trade and may have been effective in alerting miners.[53] The company also promoted safety by pitting crews of various shift bosses against each other and publishing statistics on the record of each. In a No Accident drive in 1921, thirty-nine foremen and shift bosses reported no accidents for the month of March. Overall, the accidents were reduced from fifty-six in January to forty-three in March. None of the March accidents was serious.[54]

The Homestake made a real effort to remove the obvious hazards in the mine. All dangerous places and mechanical appliances were safeguarded underground and on the surface works. Goggles were furnished at company expense to men working under conditions dangerous to the eyes. First aid materials were placed at convenient spots in the mine.[55] The staff of the United States Bureau of Mines Car No. 5 trained the men in mine rescue work and apparatus, and Homestake rescue teams competed in Bureau meets.[56] Local meets were held with cash prizes of $150 for the winning team.[57] These programs expanded in the 1930s. By 1939 prize-winning crews were given dinners in their honor, where fancy watch fobs were awarded for individual safety records.[58]

Many technical innovations by the company improved the lot of the men in the mine. Sanitary drinking fountains were installed underground, as were special latrines. Clean, well-ventilated change rooms provided the men going on and off shift with tiled showers and individual lockers.[59] Their use was required, not only for the worker's own benefit but also to avoid high-grading. The use of wet drilling methods in the 1920s reduced the dust in the air, as did the introduction of forced air ventilation in 1923.[60] Improved methods of blasting, mucking, and milling reduced the backbreaking work load. All of this was good mining technique and good business as well as being humanitarian.

World War I caused a labor shortage in the mining industry of the United States that lasted from 1917 to 1921. This lack of manpower plus the inflated cost of materials used nearly ruined the mining industry in the Black Hills. By 1920 only the Homestake and the Trojan mines were left of the twelve mines operative in 1916.[61] Conditions were so bad, in fact, that the Homestake finally had to raise wages. On March 1, 1920, a new pay scale went into effect. Underground miners got $5 per day, open-cut miners $4.75, surface miners $4.50, muckers $4.50, underground laborers $4–$4.25, and surface laborers $3.75. Other surface employees received increases of 25–50 cents per day.[62] The Homestake, always the bellwether of Black Hills mining, triggered wage raises in the other mines in the Hills.[63] By 1921, however, conditions returned to normal, and the company had second thoughts. Superintendent Yates announced a wage reduction of 50 cents per day to go into effect July 10, 1921.[64] He explained that the action was taken much later than by most other companies such as Anaconda, American Express, and U.S. Steel—which was true enough—and hoped that the employees would continue to give a "full measure of service."[65]

Of more importance to both worker and company than these pay fluctuations was the introduction of a system of contract mining in the 1920s. Under this system the company contracted with three-man teams of miners for the work they were to do. The miners agreed to do work at an agreed-upon price per unit. In this manner all development work was done on a footage basis. Drifting was paid per running foot, as was track laying. The miners were guaranteed the minimum day's wage and usually could make at least twice that amount if conditions were right and the contract was a good one. Under the system the company furnished all the equipment and tools, and the miners had to buy the explosives. The system helped the man who would work hard and efficiently. It benefited the company by increasing production and providing an incentive for faster and more efficient operations.[66] In effect, the Homestake created a group of employees who were minor capitalists. Their interests usually coincided with those of the company, and labor troubles remained at a minimum.

There were complaints, of course, under the contract system. The majority of these centered on contract pricing and the grading of the rock to be moved. The company graded the ore for each contract A, B, or C and paid according to the amount of each grade moved. In order to keep unrest minimal, the rock was regraded once a month and the grades would be shifted upward but never down if the material warranted it.[67] The contract miners were controlled by the shift bosses. The technique of giving suspensions for breaches of discipline really hurt the contract miner. He usually received these "days" for breaches of safety rule observance. The contractors cut corners to speed production if they were not carefully supervised.[68]

Contracting was not the total answer to the Homestake's labor relations. Only half of the underground miners were under contract. New men worked at day's pay until they could get on a contract. The system was never used in the surface workings and probably would have been inapplicable there. Even with these disadvantages, the contract mining system was obviously favorable to both company and worker.

The 1930s were the golden years for the Homestake and its employees. The Great Depression brought suffering to the rest of the country; but Lead, its citizens, and the company that was its sole industry boomed. Prices were as low as elsewhere in the nation while the price of gold rose by law to $35 per ounce. Wages not only held but indeed were raised in Lead. Unemployed job seekers came to the Homestake hoping to obtain one of its desirable jobs.[69] New homes and business establishments rose along the newly paved streets, and the school district constructed the best high school plant in the state.[70] Welfare continued and expanded. An extensive scholarship program for employees' college-bound children was instituted.[71] When one adds to this the welfare given by the federal government under the New Deal, it becomes obvious how well-off the people were. By 1940 the boom was leveling off as World War II grew in intensity. One of the casualties of the Japanese bombing of Pearl Harbor and America's subsequent entry into the war in 1941 was the gold mining industry. Order L-208 of the War Production Board suspended gold mining in the United States as of October 8, 1942. A period ended for the Homestake and for Lead. The underground working force, already shrunk by the draft, was dissipated. The social structure of Lead was irrevocably altered. Some milling continued, using backlogs of accumulated ore, and the foundries and shops kept busy on war work. Mining would of course reopen, and Lead would regain its lost population, but the situation would never be quite the same again. The postwar years, with their constantly inflated economy, would bring new problems which in turn would require new answers.

CHAPTER NINE

CONCLUSION

THE CLOSING of gold mines by government order in 1942 brought an end to an era in the relationship of the Homestake Mining Company and its working force. This relationship, based as it was on paternalism, differed from that of most precious metal mining companies in the American West. The Homestake's policies toward its workers created the best physical conditions in the industry and resulted in decades of labor peace, broken only once by a lockout unplanned by either side. This labor peace, combined with high productivity resulting in large, continuous profits, served both company and worker.

The Homestake Mining Company was the leader of the gold mining industry in the years from 1877 to 1942. It faced the technical problems of mining a huge body of low-grade ore and, by astute management and superior organizational skill, learned to rationalize and mechanize the rather ad hoc mining methods in use at its inception into a finely tuned method of great efficiency. In so doing it led the industry. Generally it did not pioneer new ways of doing things but rather adopted methods proven elsewhere by those with the daring and skill to try the unexplored. Others might try and fail. The Homestake copied only those who tried and succeeded. The peculiar genius of the company was in blending and organizing others' ideas and then applying them to the Homestake's situation with a keen eye for the pragmatic result. In so doing the company was always willing to change. Thus the "Homestake System" might very well mean something entirely different in 1940 than it had in 1900, but it aways meant large-scale organization of the mining and milling processes and notably fine balance sheets.

The Homestake possessed certain advantages in its struggle to master gold mining in changing eras. It was always certain of sufficient capital. The money was there to hire the right men and to buy the necessary machinery. Its owners were willing to forego the immediate coup for steady, long-haul profits that waiting and reinvesting the early profits might bring. This was in sharp contrast with the usual procedures in the business. The result was a mine that continued through decades when apparently richer lodes failed. By 1942 no mine in North America remotely compared to the Homestake. Much of this was due to wise management, but that great ore body was the basis of it all. George Hearst knew this very well; his favorite toast was "Here's to low-grade ore and plenty of it."[1]

The mining of that lode in an obscure corner of the western frontier did not allow the company to concentrate solely on the immediate productive processes. Legal peculiarities necessitated the company's ownership of the land occupied by the city housing its workers. This ownership required a peculiar system of licensing the right to occupy real property that gave the company enormous power over its workers and the city of Lead. Although the use of the power was greatly restrained, Lead was a company town with a company store and company domination. The company provided the workers with their water, their electricity, and a guarantee of their credit. Yet this control was somewhat ameliorated by the presence of competition for the worker's business and by company policy which strove to avoid exploiting the workers. The Homestake desired hard and efficient work from its people. It did not seek to rob them of their paychecks when the work was done. It did, however, want their votes and usually got them. The company's political dominance over the town and the working force was evident from the beginning, and while the crude methods of an earlier day evolved into very subtle ones later, the workers clearly voted as expected.

Working for the Homestake and living in the company town had definite advantages. The entering worker found a ready-made system of housing, home ownership, and credit awaiting him. The only necessary qualification was the company employment card. From age four his children went to the best schools available. Recreation was his for the taking. Medical service came with the job, both for the worker and for his family. The company established its own welfare state, and it took care of its members well. Immigrant workers found a warm welcome from both the company and fellow countrymen who had preceded them. They were drawn into groups of their own and then drawn out again as the company and the community force-fed them Americanism. The whole business was done without any of the ugly incidents of racial prejudice that would not have been uncommon in the United States. Racial prejudice may have existed, but not overtly; racial identification, however, was strong.

The worker at Homestake received much and no doubt deserved much. He worked hard, long, and faithfully. Yet he gave up something when he worked for the company. After the lockout his freedom to join in free association with his fellows in a union was denied him—not to be returned until the 1930s. Fundamentally, the decisions concerning his employment and his welfare determined by that employment would be made without him. His only decision could be to either retain or reject his employment with the company; frequently his life would be so entangled in problems concerning credit, home ownership, and the like that even this decision was foreordained. This is not to argue that the decisions made about the workers were necessarily bad ones. The Homestake made very good decisions for the most part. The worker was treated very well. Even so, a choice was denied. Thus employment at the Homestake was a two-edged sword that cut in more than one direction. The worker stood alone before the power of the company, albeit that it was a benevolent company. The worker had to decide for himself whether employment was worth the sacrifice—whether on balance he was ahead. Usually he decided that he was.

There can be little doubt that the Homestake system of labor relationships was a success from the company's viewpoint. The company achieved an extraordinary degree of labor peace, marred only by the 1909–10 lockout. The stability of the labor force was remarkable. In 1939, the last year before the war and the draft had their impact, only 54 men from a working force of over 2,000 quit the employ of the company.[2] In the same period 37 men left the Homestake for involuntary reasons; 9 were fired, 4 died, and the rest were laid off because of lack of work mainly in the lumbering department. An additional 103 men stopped their employment in order to return to school,[3] but they were temporary help hired for the summer and engaged mainly in the upkeep of company-owned mining claims primarily as a form of aid to education. These figures are typical of the company in the 1930s. High and efficient production in the mines and mills plus numerous employees who owned their own homes and gave every evidence of being solid, responsible community citizens were testimony that the method worked.

The company bias against unions continued and was always capable of keeping them out. Even in the 1930s when such New Deal legislation as the Wagner Labor Relations Act made overt resistance to unionism impossible, the Homestake remained an open shop. There is no evidence that organized labor even attempted to organize the camp.[4] Certainly no elections were held. One may conjecture that a view of the conditions convinced the union men that they would do well to go to more fertile fields for organization. It may be argued that this was primarily due to the unique prosperity of gold mining in the depression, and this may be somewhat true. Yet these arguments may credit the unions with more selectivity than they possessed. Certainly the Homestake would have been a feather in the cap of any labor organization, and it seems likely that they would have fought hard for it if there were any chance of success. In any case, the company retained the open shop, and methods of caring for its workers must be considered to be a major factor in this.

The company achieved its desired ends at a surprisingly small cost. In 1939, the year of greatest expenditure, the company spent $87,693.42 on the hospital program, $18,884.68 in Workmen's Compensation, $3,489.36 in contributions to the Aid Fund, $33,082.46 in benefices, $40,400.94 in pensions, and 29,808.31 on the recreation program—a total of $213,309.17.[5] This seems very small when one considers that the company produced over $19 million in gold that year, nearly one-half of which was profit.[6] Even when one takes into account the capital expenditures—recreation building, occupancy rights to surface lands, guaranteeing of loans, and all the rest of the welfare—the cost seems picayune.

The experience of the Homestake Mining Company indicated that decent treatment of the laboring force was not terribly expensive and paid the employer dividends far in excess of any costs that it entailed. Sound management could bring stability to an industry noted for opposite tendencies. It is to be wondered why others did not do as much or more—not for humanitarian reasons, but because it was so obviously good business. John R. Commons, chairman of the Commission on Industrial Relations, said at the close of its hearings, "You have here the most remarkable business organization that I have come across in the country. You have developed welfare features which are beyond anything that I know of, and they are given with a liberal hand."[7]

Any final judgment on the Homestake system of industrial relations depends to a great degree on the conditioned viewpoint of those making the judgment. Almost all agree that physically the worker was very well cared for. Bishop Joseph F. Busch, a bitter enemy of Superintendent Grier and the Homestake, stated the situation very well in the period immediately following the lockout:

> I feel that the conditions in Lead are way beyond anything that exist in any similar industrial center. . . . I have also felt that the influence of this corporation was so overwhelming in this community that if it was not used in the proper way, either consciously or unconsciously, there was much danger of a great deal of harm being done. . . . This company is so influential that any stand it takes on any question, whether it is political, moral, or religious, has a great weight with a large number of men.[8]

The *Engineering and Mining Journal* spoke glowingly of the physical conditions in Lead but said that in regard to the worker, "the only condition is that he shall have nothing to say about it himself."[9] The question had not changed by 1942, except that the worker had the right by law to say something about it himself and had not used the right. It would perhaps be impertinent to question his decision.

NOTES

INTRODUCTION

1. Peter F. Drucker, *The Concept of the Corporation*, p. 136.

CHAPTER ONE / GOLD RUSH IN THE BLACK HILLS

1. G. W. Kingsbury, *History of Dakota Territory*, I:861.
2. T. A. Rickard, *A History of American Mining*, p. 19.
3. Harold E. Briggs, "The Settlement and Economic Development of the Territory of Dakota," *South Dakota Historical Review*, I:164.
4. Weston Arthur Goodspeed, *The Province and the States*, II:271.
5. Doane Robinson, "Black Hills Bygones," *South Dakota Historical Collections*, XII:201. The stone may be viewed in the Adams Museum, Deadwood, S. Dak.
6. O. W. Coursey, *The Beautiful Black Hills*, p. 9.
7. *Black Hills Pioneer*, I:2 (June 22, 1876).
8. *Telegraph Herald*, May 5, 1878, p. 2.
9. Robert J. Casey, *The Black Hills and Their Incredible Characters*, p. 120.
10. *The Story of Homestake*, p. 5.
11. Casey, p. 120.
12. Hyman Palais, "A Survey of Early Black Hills History," *Black Hills Engineer*, XXVII:5.
13. Horatio N. MaGuire, *The Coming Empire, 1878*, quoted in Coursey, pp. 33–36.
14. *Telegraph Herald*, May 5, 1878, p. 2.
15. Goodspeed, p. 271.
16. Kingsbury, pp. 866–68.
17. Briggs, p. 164.
18. Robinson, pp. 204–5.
19. Herbert S. Schell, *History of South Dakota*, pp. 88–89.
20. Glenn Chesney Quiett, *Pay Dirt*, p. 236.
21. Charles Collins, *Black Hills History and Directory*, p. 7.
22. Quiett, p. 237.
23. Collins, p. 9.
24. Ibid., p. 11.
25. Goodspeed, p. 273.
26. R. E. Driscoll, Sr., *Seventy Years of Banking in the Black Hills*, p. 12. Also see Howard Roberts Lamar, *Dakota Territory, 1861–1889*, p. 191; Edward C. Kirkland, *A History of American Economic Life*, p. 464.
27. Rodman Wilson Paul, *Mining Frontiers of the Far West, 1848–1880*, p. 177.
28. C. C. O'Harra, "Custer's Black Hills Expedition of 1874," *Black Hills Engineer*, XVII:228.
29. Donald Jackson, *Custer's Gold; The United States Cavalry Expedition of 1874*, passim.

30. C. C. O'Harra, "Early Placer Mining in the Black Hills," *Black Hills Engineer*, XIX:343.
31. Ibid.
32. Ibid., pp. 343–44.
33. Driscoll, p. 12.
34. William Maxwell Blackburn, "Historical Sketch of North and South Dakota, 1893," *South Dakota Historical Collections*, I:64.
35. O'Harra, "Early Placer Mining," pp. 347–48.
36. Ibid.
37. Ibid.
38. Francis Church Lincoln, "Half a Century of Mining in the Black Hills," *Engineering and Mining Journal*, CXXII:205.
39. W. P. Jenney and Henry Newton, *Report on the Resources of the Black Hills of Dakota*, pp. 225–28.
40. Collins, pp. 22–23.
41. Howard Roberts Lamar, *Dakota Territory, 1861–1889*, p. 150.
42. James D. Richardson, *A Compilation of the Messages and Papers of the Presidents*, VI:4355.
43. Driscoll, p. 12.
44. Ibid.
45. O'Harra, "Early Placer Mining," p. 353.
46. R. V. Hunkins, "The Black Hills—A Storehouse of Mineral Treasure," in Roderick Peattie (ed.), *The Black Hills*, pp. 250–51.
47. George P. Baldwin, *Black Hills Illustrated*, p. 41.
48. The best description of placer mining extant is in Rodman W. Paul, *California Gold*. For Black Hills placer mining see previously cited article by C. C. O'Harra, late president of the South Dakota School of Mines and Technology.
49. Peter Rosen, *Pa-ha-sa-pah, or the Black Hills of South Dakota*, p. 397.
50. John S. McClintock, *Pioneer Days in the Black Hills*, p. 33.
51. A. D. Tallent, *The Black Hills: or The Last Hunting Ground of the Dakotahs*, pp. 171–72.
52. Muriel Sibell Wolle, *The Bonanza Trail*, p. 454.
53. O'Harra, "Early Placer Mining," p. 355.
54. *Lead Daily Call*, Jan. 1, 1905, p. 1.
55. Ibid. The figures here are based on estimates and are probably not too reliable. They do, however, give an indication of the impact that news of a gold strike had.
56. O'Harra, "Early Placer Mining," p. 355.
57. Ibid.
58. Bruce Nelson, *Land of the Dacotahs*, p. 158.
59. Doane Robinson, *A Brief History of South Dakota*, p. 153.
60. Briggs, p. 158.
61. Lincoln, p. 206.
62. "A Trip to the Black Hills," *Scribner's Monthly*, XIII:756.
63. Collins, p. 26.
64. "A Trip to the Black Hills," p. 756.
65. Ibid.

CHAPTER TWO / **HOMESTAKE**

1. Moses Manuel, untitled and unpublished manuscript. See also Rodman Wilson Paul, *Mining Frontiers of the Far West, 1848–1880*, pp. 183–84. Professor Paul mentions two other copies of the manuscript in his footnotes.
2. Robert J. Casey, *The Black Hills and Their Incredible Characters*, pp. 213–14.
3. Emma George Myron, The History of the Homestake Mine, M.A. thesis, p. 4. Professor Paul feels that no claims were located by the Manuels until 1876, which seems doubtful (Paul, p. 184).
4. Myron, p. 4.

5. Ibid., pp. 4–5.

6. G. W. Kingsbury, *History of Dakota Territory*, III:23.

7. An arastra (the term is Mexican) is a homemade device for pulverizing ore to a workable state. It is, in effect, a huge grindstone run by either mule or water power. In this case it was water power in Kirk Creek.

8. Manuel.

9. It still shows no sign of weakening in 1972.

10. George E. Roberts, *Report of the Director of the Mint upon the Production of Precious Metals in the United States During the Calendar Year 1900*, p. 186.

11. "Lucky Homestake," *Fortune* (June, 1934), p. 172.

12. T. A. Rickard, *A History of American Mining*, pp. 215–16.

13. Ibid., p. 214.

14. Edward Hungerford, *Wells Fargo, Advancing the American Frontier*, pp. 118–19.

15. "Lucky Homestake," p. 173.

16. Rickard, p. 351.

17. A. D. Tallent, *The Black Hills: or The Last Hunting Grounds of the Dakotahs*, p. 509.

18. George W. Stokes and Howard E. Driggs, *Deadwood Gold*, p. 109.

19. Ibid., pp. 110–14.

20. Tallent, p. 509.

21. Ibid.

22. *The Homestake Mining Company Report and Statement to April 1, 1880*, p. 1. This is the official report of J. B. Haggin, President and Treasurer.

23. Ibid., pp. 1–3.

24. Ibid.

25. "Lucky Homestake," p. 174.

26. Stokes and Driggs, pp. 120–21, claim that Bill Farish was the first. The Homestake, however, says that he was merely a foreman and that McMaster was the first true superintendent. See *Deadwood Pioneer Times*, Aug. 29, 1936, p. 4. Also see *The Homestake Story*, p. 5. The first annual report of J. B. Haggin also bears out the McMaster claim.

27. John S. McClintock, *Pioneer Days in the Black Hills*, p. 296.

28. Myron, p. 14.

29. Peter Rosen, *Pa-ha-sa-pah, or the Black Hills of South Dakota*, p. 430.

30. Stokes and Driggs, pp. 121, 124–25.

31. B. C. Yates, "The Homestake Mine," *Black Hills Engineer*, XIV:133.

32. Ibid.

33. Weston Arthur Goodspeed, *The Province and the States*, II:251.

34. R. V. Hunkins, "America's Greatest Gold Mine—The Homestake," in Roderick Peattie (ed.), *The Black Hills*, pp. 284–85.

35. *Lead Daily Call*, May 17, 1953.

36. Marjorie Yates Price, personal interview, Sept. 1, 1958. Mrs. Price, the daughter of Bruce Yates, has lived in Lead for years in close relationship to the Homestake management.

37. *Lead Daily Call*, May 17, 1953.

38. He had many articles in the *Engineering and Mining Journal*.

39. *Lead Daily Call*, May 17, 1953.

40. Ibid.

41. *The Story of Homestake*, pp. 7–8.

42. See Homestake Mining Company, *Annual Reports, 1878–1892*.

43. Homestake Mining Company, *Annual Report, 1892–1893*, p. 7.

44. Homestake Mining Company, *Annual Report, 1893–1894*, p. 9.

45. Homestake Mining Company, *Annual Report, 1895–1896*, p. 8.

46. Homestake Mining Company, *Annual Report, 1899–1900*.

47. Ibid., p. 11.

48. Francis Church Lincoln, "Half a Century of Mining in the Black Hills," *Engineering and Mining Journal*, CXXII:207.

49. Stokes and Driggs, p. 127.

50. Ibid.
51. Ibid.
52. Hunkins, pp. 262–63.
53. Stokes and Driggs, pp. 131–32.
54. Hunkins, pp. 262–63.
55. Frank Hebert, *Forty Years of Mining and Prospecting in the Black Hills*, p. 91.
56. Hunkins, pp. 262–63.
57. Stokes and Driggs, p. 132.
58. Ibid.
59. Casey, pp. 216–17.
60. Howard Roberts Lamar, *Dakota Territory, 1861–1889*, p. 171. Moody was a Republican and Hearst a Democrat, however.
61. Stokes and Driggs, p. 148.
62. Homestake Mining Company, *Annual Report, 1901–1902*, p. 6.
63. Ibid.
64. Lincoln, p. 207.
65. Ibid.
66. Homestake Mining Company, *Annual Report, 1902–1903*, pp. 1–8.
67. Homestake Mining Company, *Annual Report, 1903–1904*, p. 4.
68. Homestake Mining Company, *Annual Report, 1905–1906*, p. 7.
69. *Lead Daily Call*, May 17, 1953.
70. Kenneth C. Kellar, personal interview, Dec. 29, 1964. Mr. Kellar is vice-president and chief legal counsel, and son of the former chief legal counsel of the Homestake.
71. Homestake Mining Company, *Annual Reports, 1917–1920*.
72. Homestake Mining Company, *Annual Report, 1923*.
73. Dr. McLaughlin would become a world-famous figure, serving as consultant to many mining companies. He also is a distinguished scholar and educator, heading the Geology and Geography Department at Harvard, serving as Dean of the College of Mining at the University of California and Dean of the College of Engineering at the same institution. After World War II he became president of the Homestake and at the present time is chairman of the board. See *Lead Daily Call*, May 17, 1953.
74. D. P. Howe, personal interview, Jan. 2, 1965. Howe is the Homestake public relations director, editor of *Sharp Bits*, and a published author of Black Hills history.
75. Homestake Mining Company, *Annual Report, 1931*.
76. Homestake Mining Company, *Annual Report, 1935*.
77. Clarence N. Kravig, "Mining in Lawrence County," in Mildred Fielder (ed.), *Lawrence County*, p. 98. Kravig is the present assistant manager of the Homestake.

CHAPTER THREE / **THE WORKERS**

1. The dangers of mining are plentiful under the best of conditions. Mistakes in timbering, barring down, or using explosives can be fatal to a whole crew.
2. See R. Page Arnot, *The Miners*.
3. The only work covering labor relations in this field is entitled *Heritage of Conflict*. In it Professor Jensen of Cornell saw "the cankering effect which metals have had upon men," and while he sees heroic men, there are "those molded, not to say deformed, by the earth's wickedness and hardness." While this sounds like a perverse form of chemistry worked by the rocks upon man, Jensen does see labor relations as a "product of environmental factors" (V. H. Jensen, *Heritage of Conflict*, passim).
4. It has proven impossible to document this, but there seems to have been no alternate labor supply, and it was the normal procedure in most rushes. See Jensen, p. 4.
5. Rodman Wilson Paul, *Mining Frontiers of the Far West, 1848–1880*,

passim. Professor Paul correctly emphasizes the impact of the Cornish on the mining frontier in general.

6. Joseph H. Cash, A History of Lead, South Dakota, 1876–1900, M.A. thesis, pp. 118–21.

7. Jelbert Morcom, personal interview, Aug. 13, 1958. Mr. Morcom is a Cousin Jack who is much interested in his fellows.

8. J. A. Jobe, personal interview, Oct. 8, 1958. Jobe was born in Cornwall and learned mining there. He migrated to Michigan and from there to Lead in 1902 where he worked as a hoist operator and was a union member until the lockout. His voluminous notes, extensive clipping collections, and remarkably accurate memory made him the perfect interviewee.

9. United States Bureau of Census, *Tenth Census: 1880,* pp. 735, 814. The figures here are for Dakota Territory as a whole but in reality apply to Black Hills mining. It is assumed that the Homestake reflects these. The medical logs of Dr. J. W. Freeman, who worked as contract surgeon for the company until he became chief surgeon in 1904, indicate that his patients were mainly English, Scotch, and Irish in descent until the mid-1880s. There were also some Germans and Scandinavians.

10. Helen Morganti, personal interview, Sept. 4, 1958. Miss Morganti has published on Black Hills history, speaks Italian fluently, and is very closely connected to the Italian community in Lead.

11. Margaret Stabio, personal interview, Sept. 22, 1958. Miss Stabio has lived her entire life in Lead and has studied her fellow Italians closely.

12. Jerry Batinovich, personal interview, Oct. 6, 1958. Practically nothing has been published on the Slavonians in Lead, but Batinovich is one and has studied them and may be considered an expert on the subject. He has a private museum of Slavonian artifacts.

13. John Mastrovich, personal interview, Mar. 18, 1958. Mastrovich, a pro-union barber, has lived in Slavonian Alley for over six decades.

14. Morganti.

15. Batinovich.

16. Herbert S. Schell, *History of South Dakota,* p. 115.

17. A. D. Tallent, *The Black Hills: or The Last Hunting Ground of the Dakotahs,* pp. 526–27.

18. Just why this is so is hard to determine. W. R. Hearst was always worried about a "yellow menace," but he had nothing to do with the Homestake, and his mother gives no evidence of prejudice. It is possible that none tried for jobs in mining.

19. Francis Church Lincoln, "Half a Century of Mining in the Black Hills," *Engineering and Mining Journal,* CXXII: 214.

20. Ibid.

21. Jobe.

22. Clarence Kravig, personal interview, Jan. 5, 1965.

23. Ibid.

24. Ibid.

25. This is difficult to prove. Many interviewed, if not all, claim it was practiced by shift bosses.

26. Kragiv.

27. *Lead Daily Call,* Mar. 12, 1901.

28. Paul, p. 182.

29. Homestake Mining Company, *Annual Report, Jan. 1, 1878–Sept. 1, 1880.* This first corporate report includes McMaster's original report. Paul's work concludes in 1880 and so would coincide in time with it.

30. *Lead Daily Call,* July 24, 1905.

31. T. A. Rickard, *The Stamp Milling of Gold Ores,* pp. 220–21.

32. Ibid., p. 96.

33. *Lead Daily Call,* Apr. 14, 1906.

34. Clarence N. Kravig, "Mining in Lawrence County," in Mildred Fielder (ed.), *Lawrence County,* p. 98.

35. *Lead Daily Call,* Jan. 12, 1900.

36. *Lead Daily Call*, Mar. 17, 1902.
37. See Chapter 2.
38. Tallent, p. 520.
39. Peter Rosen, *Pa-ha-sa-pah, or the Black Hills of South Dakota*, p. 428.
40. The Court of the 12th Judicial Circuit, State of South Dakota, confirmed its charitable nature in 1899 and exempted it from taxation. *Lead Daily Call*, Dec. 28, 1899.
41. *Lead City Miners' Union Minute Book, 1890–1901*, p. 10.
42. Ibid., p. 66.
43. Ibid., pp. 149–50.
44. Tallent, pp. 520–21.
45. *Lead City Miners' Union Minute Book*, pp. 149–50.
46. Ibid., passim.
47. Ibid., p. 164.
48. Ibid., p. 337.
49. Ibid., p. 211.
50. U.S. Commission on Industrial Relations, *Final Report and Testimony*, IV: 3563. This is on the testimony of T. J. Grier.
51. *Lead City Miners' Union Minute Book*, p. 83.
52. Ibid.
53. Ibid., p. 204.
54. Tallent, p. 523.
55. George P. Baldwin, *Black Hills Illustrated*, p. 105.
56. *The Lantern*, Dec. 16, 1909, p. 4.
57. *Evening Call*, Mar. 13, 1901.
58. *Miners' Magazine*, Nov. 1902, p. 35.
59. U.S. Commission on Industrial Relations, IV:3566–67.
60. Jobe.
61. U.S. Commission on Industrial Relations, IV:3603.
62. W. C. Benfer, "The Story of the Homestake Lockout," *International Socialist Review*, X:782.
63. *Lead Daily Call*, Dec. 13, 1906. It must be admitted that the *Daily Call* tended to reflect the company attitude.

CHAPTER FOUR / MINING AND MILLING

1. Rodman Wilson Paul, *Mining Frontiers of the Far West, 1848–1880*, p. 7.
2. This is well illustrated by the simple fact that the Homestake, whose ore is not at all rich, is the only major gold mine still operative in the United States.
3. T. A. Rickard, *A History of American Mining*, p. 217.
4. B. C. Yates, "Some Features of Mining Operations in the Homestake Mine," *Black Hills Illustrated*, p. 14.
5. Pay shoot: that portion of a vein which carries the profitable or pay ore (Albert H. Fay, *A Glossary of the Mining and Mineral Industry*).
6. *The Homestake Story*, p. 9.
7. Rickard, p. 217.
8. The Homestake Mining Company, *Report and Statement to April 1, 1880*.
9. Chute: a channel or shaft underground, or an inclined trough above ground, through which ore falls or is "shot" by gravity from a higher to a lower level (Fay).
10. Emma George Myron. The History of the Homestake Mine, M.A. thesis, p. 23.
11. Paul, p. 182.
12. Ibid., p. 64.
13. Myron, p. 24.
14. A. J. M. Ross and R. G. Wayland, "A Brief Outline of Homestake Mining Methods," *Black Hills Engineer*, XIV: 145.

15. Myron, p. 25.
16. Ross and Wayland, p. 147.
17. Myron, pp. 25–27.
18. Ross and Wayland, p. 147.
19. *Sharp Bits*, IV:2 (Apr. 1953).
20. Ibid.
21. Clarence N. Kravig, "Mining in Lawrence County," in Mildred Fielder (ed.), *Lawrence County*, p. 98.
22. *Sharp Bits*, IV:2 (Apr. 1953).
23. Ibid., p. 1.
24. Ross and Wayland, p. 153.
25. Paul, p. 94.
26. *Sharp Bits*, IV:4 (Apr. 1953).
27. Homestake Mining Company, *Annual Report, 1894–1895*.
28. Kravig, p. 99.
29. Ibid., p. 98.
30. Ross and Wayland, p. 147.
31. Francis Church Lincoln, "Half a Century of Mining in the Black Hills," *Engineering and Mining Journal*, CXXII:211.
32. Lincoln, pp. 211–12.
33. Myron, p. 31.
34. J. A. Jobe, personal interview, Oct. 8, 1958.
35. Lincoln, p. 212.
36. *Sharp Bits*, IX:2 (Oct. 1958).
37. Myron, p. 30.
38. Ibid., p. 31.
39. Skip: a large hoisting bucket constructed of boiler plate, which slides between guides in a shaft. The bail is usually connected at or near bottom of bucket so that it may be automatically dumped at the surface (Fay).
40. Lincoln, p. 212.
41. Myron, p. 32.
42. Lincoln, p. 212.
43. Myron, p. 32.
44. *Lead Daily Call*, May 17, 1953.
45. *Sharp Bits*, IV:1 (June 1953).
46. T. A. Rickard, *The Stamp Milling of Gold Ores*, p. 1.
47. Ibid., p. 2.
48. Rickard, *History of American Mining*, p. 217.
49. Lincoln, p. 213.
50. Rickard, *History of American Mining*, p. 217.
51. Myron, p. 35.
52. *Lead Daily Call*, May 17, 1953.
53. Myron, p. 36.
54. *Lead Daily Call*, May 17, 1953.
55. Rickard, *History of American Mining*, p. 217.
56. Rickard, *Stamp Milling of Gold Ores*, p. 99.
57. *Lead Daily Call*, May 17, 1953.
58. Lincoln, p. 213.
59. Rickard, *History of American Mining*, p. 220.
60. *Lead Daily Call*, May 17, 1953.
61. A. D. Tallent, *The Black Hills: or The Last Hunting Grounds of the Dakotahs*, p. 297.
62. Myron, p. 37.
63. *Lead Daily Call*, May 17, 1953.
64. Ibid.
65. George P. Baldwin, *Black Hills Illustrated*, p. 125.
66. Lincoln, p. 213.
67. *Lead Daily Call*, May 17, 1953.
68. Lincoln, p. 213.

69. Ibid.
70. Rickard, *History of American Mining*, p. 220.
71. *Sharp Bits*, VIII:6 (July 1957).
72. Myron, p. 40.
73. Ibid., p. 42.
74. *Sharp Bits*, VII:4 (May 1956).
75. *Lead Daily Call*, May 17, 1953; Myron, pp. 42–43.
76. *Lead Daily Call*, May 17, 1953.
77. Ibid.
78. Ibid.
79. Baldwin, p. 126. Grier stated that there was 20 years' ore in sight in 1904. Usually they tried to keep 12 years' ore in sight.

CHAPTER FIVE / **THE COMPANY TOWN**

1. Rodman Wilson Paul, *Mining Frontiers of the Far West, 1848–1880*, p. 9.
2. Ibid., p. 196.
3. The town was originally named Lead City. In order to avoid confusion, Lead will be used herein unless the context demands the older term.
4. James Brown Allen, The Company Town as a Feature of Western American Development, Ph.D. dissertation. This study does not include Lead.
5. Ibid., pp. 194–95.
6. Ibid., p. 256.
7. Jesse Brown and A. M. Willard, *Black Hills Trails*, pp. 473–74.
8. Thomas F. Carey, recorder, *Gold Run First Record Book*, p. 145.
9. Ibid., pp. 143–54.
10. See Donald P. Howe, "Lead," in Mildred Fielder (ed.), *Lawrence County*, pp. 31–32. Howe, the able publicist for the Homestake, concludes that Lead City was founded first and then Washington. He bases this on a 1906 sketch by W. P. Raddick, a gold rush pioneer. I have differed with this view in the past but am willing to concede that there is no positive documented proof that can settle the question. See Joseph H. Cash, A History of Lead, South Dakota, 1876–1900, M.A. thesis, pp. 20–22. This view is based on an eyewitness account that I consider superior to Raddick.
11. Richard B. Hughes, *Pioneer Years in the Black Hills*, p. 202. Prospectors were seldom clean, but Smokey's filth was a legend. On the one recorded case of his cleaning up, no one recognized him.
12. Brown and Willard, pp. 473–74.
13. W. P. Raddick, "Early History of Lead," *Lead Daily Call*, p. 1. From casual observance it seems that Smokey's pocket compass did as capable a job as Von Bodingen's transit, chain, and rod.
14. Hughes, p. 202.
15. Raddick, p. 1.
16. Ibid.
17. Ibid.
18. Ibid.
19. Richard Furze, personal interview, Dec. 16, 1958. Judge Furze is the municipal judge of Lead and has thoroughly studied all phases of property ownership in the city.
20. *Illustrated Booster Nugget*.
21. Furze.
22. Ibid.
23. Ibid.
24. George P. Baldwin, *Black Hills Illustrated*, p. 128.
25. Furze.
26. Baldwin, p. 128.
27. The original trustees, on their deaths, were not replaced. The incumbent county judge of Lawrence County continues to issue deeds when necessary (Furze).

28. Furze.
29. Ibid.
30. Kenneth Kellar, personal interview, Dec. 29, 1964.
31. Raddick, p. 1.
32. Brown and Willard, p. 473.
33. Muriel Sibell Wolle, *The Bonanza Trail*, p. 474.
34. Walter Shelley Phillips, *The Old-Timer's Tale*, p. 31.
35. McClure, "Dakota," *Harpers New Monthly Magazine*, XVIII:351. The roar did not cease until the 1950s when efficient quiet devices replaced the last of the stamps.
36. Both the reminiscences of early settlers and photographs of the city confirm this impression.
37. Mrs. Horace Clark, personal interview, Dec. 10, 1958. Mrs. Clark came to Lead in the 1890s and was a constant collector of material on the city.
38. *Lead Daily Call*, May 17, 1953.
39. Ibid. The area still is settling slightly, but the major portion ended by 1939.
40. Ibid.
41. J. A. Jobe, personal interview, Oct. 8, 1958.
42. Charles Collins, *Black Hills History and Directory*, p. 83.
43. *Lead City Daily Tribune*, Aug. 24, 1882.
44. Willis E. Johnson, *South Dakota, A Republic of Friends*, p. 320.
45. United States Bureau of the Census, *Twelfth Census: 1900.*
46. *Lead Daily Call*, Aug. 5, 1926, p. 18.
47. Ibid.
48. Ibid.
49. Oliver Carlson and Ernest Sutherland Bates, *Hearst, Lord of San Simeon*, p. 297.
50. The Homestake paid in gold as long as there was a demand for it and as long as it was legal.
51. If a man lived until 1943, when fire consumed the store and all its records, he was cleared of his bill. The fire served the men better than it did the Hearsts and the historian. Most of the information on the store must come from conversation with its former customers and employees, notably Everett Cotton, former manager of the hardware department.
52. Agnes Wright Spring, *The Cheyenne and Black Hills Stage and Express Routes*, p. 295.
53. R. E. Driscoll, Sr., *Seventy Years of Banking in the Black Hills*, passim.
54. A. D. Tallent, *The Black Hills: or The Last Hunting Grounds of the Dakotahs*, p. 525.
55. Driscoll, p. 30.
56. Ibid., p. 31.
57. Ibid.
58. It is not unheard of for an employee to have his house loan at the First National and his checking account at a competing bank.
59. See Chapter 2.
60. Howard Roberts Lamar, *Dakota Territory, 1861–1889*, p. 171.
61. R. V. Hunkins, personal interview, Aug. 13, 1958. Hunkins was the superintendent of schools in Lead for decades and has published on the Homestake.
62. Furze.
63. Journal #1, Hearst Free Library, p. 38. The librarian liked to jot down the important events in the library journal.
64. See Chapter 7.
65. Furze; Hunkins.
66. *Sharp Bits*, June, 1958, p. 3. In the grim 1950s and 1960s, the Homestake would regret this and try desperately to either repeal or amend the tax.
67. "Butte and Lead Contrasted," *Engineering and Mining Journal*, XCVIII:378.
68. See Chapters 6 and 8.

69. See Chapter 7.
70. Paul, p. 9.

CHAPTER SIX / **WELFARE: 1877–1910**

1. *Lead Daily Call,* May 17, 1953.
2. See Chapter 3.
3. Mrs. Horace Clark, personal interview, Dec. 10, 1958.
4. *Lead City Daily Tribune,* Mar. 29, 1899.
5. George P. Baldwin, *Black Hills Illustrated,* p. 105.
6. *Lead City Miners' Union Minute Book,* p. 115.
7. See Chapter 3.
8. *Lead City Miners' Union Minute Book,* p. 387.
9. See Chapter 3.
10. *Lead City Daily Tribune,* Feb. 24, 1892.
11. Margaret Stabio, personal interview, Sept. 22, 1958.
12. J. E. Meddaugh, *Souvenir of Lead—Black Hills Metropolis,* p. 1.
13. All interviewees agreed that this was the case, whether or not the local people were forewarned of the man's coming.
14. Jerry Batinovich, personal interview, Oct. 6, 1958.
15. Ibid.; Helen Morganti, personal interview, Sept. 4, 1958.
16. Batinovich.
17. This tendency still continues, but to a much lesser degree.
18. Oliver Carlson and Ernest Sutherland Bates, *Hearst, Lord of San
Simeon,* pp. 36–37. There is no disagreement on Mrs. Hearst. Even biographers
hostile to her son, such as Carlson and Bates, admire her.
19. R. V. Hunkins, "America's Greatest Gold Mine—The Homestake," in
Roderick Peattie (ed.), *The Black Hills,* pp. 284–85.
20. This backfired during the lockout of 1909–10. William Randolph received much criticism for this, although he had nothing to do with the management of the company.
21. Journal #1, Hearst Free Library, pp. 3–5.
22. *Illustrated Booster Nugget,* pp. 78–79.
23. A. D. Tallent, *The Black Hills: or The Last Hunting Grounds of the
Dakotahs,* p. 520.
24. Ibid., p. 525.
25. Ledger #200, Hearst Free Library.
26. Tallent, p. 525.
27. B. C. Yates, "Welfare Work at the Homestake Mine," *Engineering
and Mining Journal,* CX:199.
28. Journal #1, p. 70.
29. *Lead Daily Call,* Aug. 16, 1902.
30. Ibid., May 17, 1953.
31. *Evening Call,* June 8, 1900.
32. Ibid., Aug. 17, 1900.
33. Minutes of the Board of Education, p. 97.
34. Winifred Black Bonfils, *The Life and Personality of Phoebe Apperson
Hearst,* pp. 57–58. This biography was written at the behest of William Randolph Hearst by his first "sob sister." It was beautifully printed and bound by
John Henry Nash in a limited edition of 1,000 copies. Copy #56 was presented
to the Homestake Library. A similar volume of the life of Senator George
Hearst was given to the Hearst Kindergarten. Both are examples of badly done
biography and must be used sparingly and with the greatest care.
35. *Evening Call,* Apr. 17, 1901.
36. Ibid., Apr. 1, 1904.
37. *Illustrated Booster Nugget,* p. 95.
38. Bonfils, p. 58.
39. Yates, p. 199.
40. *Evening Call,* Aug. 16, 1902. The practice was later continued by the
Homestake.

41. See Chapter 7.
42. W. A. Swanberg, *Citizen Hearst,* p. 73.
43. Ibid., p. 88.
44. Ibid., p. 382.
45. *Evening Call,* Mar. 11, 1903.
46. Ibid., Nov. 8, 1905.
47. See Chapter 5 for a complete explanation of this.
48. Baldwin, p. 128.
49. Ibid.
50. Yates, p. 198.
51. W. P. Raddick, "Early History of Lead," *Lead Daily Call,* Aug. 5, 1926, p. 2.
52. A. T. Andreas, *Historical Atlas of Dakota,* p. 213.
53. F. S. Howe, *Deadwood Doctor,* pp. 14, 43.
54. *Lead Daily Call,* May 17, 1953. This hospital was later turned into a school building. The Sisters moved to Deadwood.
55. Ibid.
56. Emma George Myron, The History of the Homestake Mine, M.A. thesis, p. 51.
57. Yates, p. 200.
58. Ibid.
59. *Lead Daily Call,* Dec. 21, 1905.
60. Yates, p. 200.
61. The annual reports of the company indicate that the payments covered the hospital costs, but not that of the doctors or drugs. With over 2,000 employees, the $1.10 added up to a considerable sum; any firm would think carefully before dispensing with the charge.
62. Yates, p. 199.
63. Raddick, p. 1.
64. *Directory,* Lead Public Schools, p. 6.
65. R. E. Driscoll, Sr., *Seventy Years of Banking in the Black Hills,* pp. 14–15.
66. Yates, p. 199.
67. Driscoll, p. 31.
68. Minutes of the Board of Education, pp. 193–94.
69. Rev. John I. Sanford, *The Black Hills Souvenir,* p. 178.
70. *Course of Study of the Public Schools of Lead, South Dakota,* pp. 5–7.
71. Minutes of the Board of Education, p. 15.
72. *Fourth Biennial Report of the Superintendent of Public Instruction,* pp. 36–40.
73. Ibid., p. 97.
74. W. F. Hall, Education in the Black Hills before 1900, M.A. thesis, p. 22.
75. *Lead Daily Call,* Mar. 29, 1906.
76. Ibid., Nov. 2, 1906.
77. Yates, p. 199.
78. *Lead Daily Call,* July 23, 1902.
79. Yates, p. 199. The present superintendent, James Harder, is the son of a former miner and worked in the pit himself.

CHAPTER SEVEN / **THE LOCKOUT**

1. William D. Haywood, *Bill Haywood's Book,* p. 280.
2. *Miners' Magazine,* X:10 (Feb. 18, 1909).
3. *Lead Daily Call,* Sept. 9, 1909, p. 9.
4. U.S. Commission on Industrial Relations, *Final Report and Testimony,* IV:3602–3.
5. *Lead Daily Call,* Oct. 11, 1909, p. 1.
6. W. C. Benfer, "The Story of the Homestake Lockout," *International Socialist Review,* X:783.

7. U.S. Commission on Industrial Relations, p. 3604.

8. *The Lantern*, Oct. 14, 1909, p. 1.

9. *Miners' Magazine*, XI:5 (Nov. 4, 1909).

10. J. A. Jobe, personal interview, May 26, 1961.

11. Benfer, p. 783.

12. U.S. Commission on Industrial Relations, p. 3565.

13. Ibid., p. 3571.

14. *Lead Daily Call*, Dec. 27, 1909, p. 2.

15. *The Lantern*, Nov. 11, 1909, p. 1.

16. *Miners' Magazine*, XI:7 (Dec. 16, 1909).

17. U.S. Commission on Industrial Relations, p. 3604.

18. Ibid., p. 3571.

19. *Lead Daily Call*, Nov. 17, 1909, p. 1.

20. Ibid., Nov. 19, 1909, p. 1; *The Lantern*, Nov. 25, 1909, p. 1; *Miners' Magazine*, XI:10 (Nov. 25, 1909).

21. *Lead Daily Call*, Nov. 1, 1909, p. 1.

22. *The Lantern*, Nov. 25, 1909, p. 1.

23. *Lead Daily Call*, Nov. 22, 1909, p. 1; U.S. Commission on Industrial Relations, p. 3567.

24. William F. Tracy, "Letter to the Editor," *Lead Daily Call*, Nov. 15, 1909, p. 2.

25. U.S. Commission on Industrial Relations, p. 3603.

26. *Lead Daily Call*, Nov. 27, 1909, p. 2.

27. Jobe.

28. *Lead Daily Call*, Nov. 24, 1909, p. 1.

29. Jobe.

30. "Work Suspended in All Departments," *Deadwood Pioneer Times*, Nov. 25, 1909, p. 1.

31. Blanche Coleman, personal interview, May 3, 1961. Miss Coleman was a legal clerk in Kellar's office during the lockout.

32. "Work Suspended," p. 1.

33. *Lead Daily Call*, Dec. 2, 1909, p. 1.

34. Benfer, p. 785.

35. *The Lantern*, Dec. 2, 1909, p. 4.

36. *Miners' Magazine*, XI:2 (Dec. 9, 1909).

37. *Lead Daily Call*, Dec. 22, 1909, p. 1.

38. Jobe.

39. Marjorie Yates Price, personal interview, Oct. 12, 1960.

40. Mrs. Horace S. Clark, personal interview, Dec. 10, 1958.

41. *Lead Daily Call*, Nov. 29, 1909, p. 1. The pro-union papers admitted the incident took place but considered it to be proper behavior under the circumstances.

42. Ibid., Dec. 3, 1909, p. 1.

43. Ibid., Dec. 4, 1909, p. 1.

44. Ibid., Dec. 15, 1909, p. 1.

45. Benfer, p. 787.

46. *Miners' Magazine*, XI:4 (Dec. 24, 1909).

47. *The Lantern*, Dec. 16, 1909, p. 1.

48. Benfer, p. 787.

49. Jobe.

50. *Miners' Magazine*, XI:4 (Apr. 21, 1910).

51. *Official Proceedings of the Nineteenth Annual Convention, Western Federation of Miners*, 1911, p. 116.

52. *Miners' Magazine*, XI:4 (Apr. 19, 1910).

53. *Official Proceedings of the Eighteenth Annual Convention, Western Federation of Miners*, 1910, p. 112.

54. *Official Proceedings, Nineteenth Annual Convention*, pp. 175, 177.

55. *Miners' Magazine*, XI:10 (Dec. 16, 1909).

56. *Lead Daily Call*, Nov. 2, 1909, p. 2.

57. *Miners' Magazine*, XI:9 (Jan. 6, 1910).
58. *The Lantern*, Aug. 12, 1909, p. 2.
59. Ibid., Feb. 10, 1910, p. 1.
60. Jobe.
61. *The Lantern*, Feb. 24, 1910, p. 1; *Miners' Magazine*, XI:7 (May 19, 1910).
62. *Lead Daily Call*, Feb. 8, 1910, p. 1.
63. Ibid., Apr. 18, 1910, p. 1.
64. Ibid., Apr. 19, 1910, p. 1.
65. *The Lantern*, Apr. 7, 1910, p. 1.
66. *Official Proceedings, Eighteenth Annual Convention*, p. 291.
67. U.S. Commission on Industrial Relations, p. 3569.
68. *Miners' Magazine*, XI:5 (Dec. 16, 1919).
69. *Lead Daily Call*, Dec. 16, 1909, p. 2.
70. Ibid.
71. Ibid., Sept. 1, 1909, p. 1.
72. *Miners' Magazine*, XI:5 (Sept. 16, 1909).
73. *Official Proceedings, Eighteenth Annual Convention*, p. 291.
74. *The Lantern*, Jan. 27, 1910, p. 3.
75. Walter Renton Ingalls, ed., *The Mineral Industry During 1909*, XVIII: 296.
76. U.S. Commission on Industrial Relations, p. 3568.
77. Ingalls, p. 296.
78. *Lead Daily Call, Supplement*, Mar. 9, 1910.
79. *Lead Daily Call*, Jan. 12, 1910, p. 1.
80. *The Lantern*, Dec. 23, 1909, p. 1.
81. Ibid., Feb. 10, 1910, p. 2.
82. *Lead Daily Call*, Mar. 31, 1910, p. 1.
83. S. Goodale Price, personal interview, Apr. 15, 1961. Mr. Price, a widely published professional writer, was a patient in the hospital at the time.
84. Kenneth Kellar, personal interview, Apr. 18, 1961.
85. Jobe.
86. *The Lantern*, Jan. 13, 1910, p. 1.
87. Ibid.
88. *Official Proceedings, Eighteenth Annual Convention*, p. 290.
89. *The Lantern*, Jan. 13, 1910, p. 4.
90. Ibid.
91. Benfer, p. 786.
92. Ibid., pp. 786–87.
93. Jobe.
94. *The Lantern*, Feb. 10, 1910; Marjorie Yates Price.
95. *Miners' Magazine*, XII:14 (Mar. 3, 1910).
96. Ibid., XII:5 (May 19, 1910).
97. *Official Proceedings, Eighteenth Annual Convention*, p. 291.
98. Marjorie Yates Price.
99. *Lead Daily Call*, Jan. 10, 1910, p. 1.
100. Jobe.
101. *Lead Daily Call*, Feb. 1, 1919, p. 1.
102. *The Lantern*, Feb. 10, 1910, p. 1.
103. Benfer, p. 787.
104. *The Lantern*, Feb. 17, 1910, p. 1.
105. *Lead Daily Call*, Mar. 3, 1910, p. 1.
106. U.S. Commission on Industrial Relations, p. 3570.
107. *The Lantern*, Mar. 3, 1910, p. 1.
108. *Miners' Magazine*, XII:4 (Mar. 3, 1910).
109. Jobe.
110. *Lead Daily Call*, Mar. 31, 1910, p. 1.
111. *Official Proceedings, Nineteenth Annual Convention*, p. 4.
112. Ibid., p. 45.

113. *Official Proceedings of the Twenty-First Annual Convention, Western Federation of Miners, 1913,* p. 4.
114. U.S. Commission on Industrial Relations, p. 3569.
115. *Official Proceedings, Nineteenth Annual Convention,* p. 5.
116. U.S. Commission on Industrial Relations, p. 3569.
117. Ibid., pp. 3619–20.
118. Ibid., p. 3621.
119. *Lead Daily Call,* Aug. 22, 1910.
120. Ibid., Nov. 28, 1910.
121. "Butte and Lead Contrasted," *Engineering and Mining Journal,* XCVIII:378.
122. "Lead—August 8," *Engineering and Mining Journal,* XCVIII:324.
123. U.S. Commission on Industrial Relations, p. 3595.
124. Ibid.

CHAPTER EIGHT / CULMINATION OF THE SYSTEM: 1910–42

1. This belief has never been flatly stated but is definitely there. Conversations with company officials show it, and company policy confirms it.
2. *Lead Daily Call,* Oct. 20, 1910.
3. J. A. Jobe, personal interview, May 26, 1961. Jobe was a patient in the hospital during the lockout.
4. B. C. Yates, "Welfare Work at the Homestake Mine," *Engineering and Mining Journal,* CX:198 (July 31, 1920).
5. Yates's statement was sound conservatism—in the Burkean sense of the term—and was based on a view of society as an organism subject to organic growth.
6. See Chapter 7.
7. Homestake Mining Company, *Annual Report, 1912.*
8. Yates, p. 200.
9. Ibid.
10. *Lead Daily Call,* Aug. 1, 1919.
11. Yates, p. 201.
12. Ibid.
13. Joe Dunmire, personal interview, Jan. 2, 1964. Dunmire is the long-time head of the Recreation Department and professionally educated in the field of recreation and physical education. The cost for a family membership at present is $39 per year. It is not unusual to see a golf foursome that includes a mechanic, a miner, the superintendent, and the chief counsel.
14. Ibid.
15. Ibid.
16. Ibid.
17. Ibid.
18. Guy N. Bjorge, History of the Homestake Employees Aid Fund, a letter to all members, June 13, 1947.
19. Francis Church Lincoln, "Half a Century of Mining in the Black Hills," *Engineering and Mining Journal,* CXXII:214.
20. *Lead Daily Call,* May 17, 1953. It must be remembered that the employees and their families had full free medical care and hospitalization.
21. Yates, p. 201.
22. *Lead Daily Call,* May 17, 1953.
23. Ibid., Dec. 18, 1918.
24. Yates, p. 201.
25. Ibid., p. 202.
26. Ibid.
27. Donald P. Howe, personal interview, Jan. 2, 1965.
28. Yates, p. 201.
29. Homestake Mining Company, *Employees Record of Underground Service.*
30. *Lead Daily Call,* Apr. 14, 1919.

31. Yates, p. 201.

32. B. C. Yates, letter to Homestake employees, Apr. 1, 1919. The letter is in the private collection of the author.

33. Homestake Mining Company, Foreign-Born—a list of workers that is undated but apparently compiled in the late 1920s. It lists approximately 700 foreign-born.

34. *Lead Daily Call*, May 17, 1953.

35. Letter from D. P. Mitchell to Emma George Myron, Oct. 21, 1931.

36. See Yates, "Welfare Work," passim.

37. *Lead Daily Call*, May 17, 1953.

38. Yates, "Welfare Work," p. 200.

39. *Lead Daily Call*, May 17, 1953.

40. Yates, "Welfare Work," p. 200. This is still the case. There are no nurse's aides or even practical nurses employed by the company hospital.

41. Homestake Mining Company, *Report of the Hospital Department,* 1924, p. 1.

42. Ibid., 1929, p. 1.

43. Ibid., 1940, p. 1.

44. Homestake Mining Company, *Report on Silicosis*, Sept. 5, 1936, passim.

45. Arthur M. Semones, personal interview, Jan. 7, 1965. Dr. Semones is the present Homestake Mining Company Chief Surgeon.

46. Ibid.

47. Ibid.

48. See Jewett V. Reed and A. K. Harcourt, *The Essentials of Occupational Diseases*, pp. 161–74, for a discussion of the disease.

49. Semones.

50. Yates, "Welfare Work," p. 203.

51. *Sharp Bits*, XIV:2 (Mar. 1963).

52. C. A. Brooks, "Accident Prevention at the Homestake Mine," *Explosives Engineer*, IV:209–12.

53. *The Homestake Safety Bulletin*, I:passim.

54. *Lead Daily Call*, Apr. 21, 1921.

55. Yates, "Welfare Work," p. 203.

56. *Engineering and Mining Journal*, CX:41 (July 3, 1920).

57. *Engineering and Mining Journal*, CX:87 (July 10, 1920).

58. *Sharp Bits*, XIII:29 (Feb. 1962).

59. Yates, "Welfare Work," p. 203.

60. Lincoln, pp. 211–12.

61. B. C. Yates, "South Dakota Gold Mining," *Engineering and Mining Journal*, CX:1169 (Dec. 18, 1920).

62. *Lead Daily Call*, Feb. 20, 1920.

63. Ibid., Mar. 4, 1920.

64. Ibid., June 25, 1921.

65. Ibid.

66. Clarence Kravig, personal interview, Jan. 5, 1965. Kravig is second in command at the Homestake at the present time. He has worked since World War I, starting as a mining surveyor.

67. Ibid.

68. Ibid.

69. Donald P. Howe, "Lead," in Mildred Fielder (ed.), *Lawrence County*, p. 42.

70. Ibid., p. 43. The Homestake bought the old school for $250,000, provided a site free, and later gave the old school back to the district.

71. *Sharp Bits*, X:passim (Aug. 1959).

CHAPTER NINE / CONCLUSION

1. *Lead Daily Call*, May 1, 1904.

2. Homestake Mining Company, *Hiring Report, 1939, Employment Department*.

3. Ibid.

4. *Lead Daily Call,* May 15, 1963.

5. Homestake Mining Company, *Annual Report, 1939.*

6. Ibid. The figures are fairly representative of the 1930s. Prior to that the expenditures are difficult to determine with exactness, but the expenditure seems similar. The price of gold was much less, however.

7. *Engineering and Mining Journal,* Aug. 8, 1914, p. 324; Oct. 2, 1914, p. 616.

8. U.S. Commission on Industrial Relations, *Final Report and Testimony,* IV:3590.

9. *Engineering and Mining Journal,* Aug. 29, 1914, p. 378.

BIBLIOGRAPHY

PRIMARY SOURCES

MANUSCRIPTS

Allen, James Brown. The Company Town as a Feature of Western Development. Unpublished Ph.D. dissertation, Univ. S. Calif., 1963.

Carey, T. F., Recorder. *Gold Run First Record Book.* Feb. 21, 1876. This may be seen at the Adams Memorial Museum, Deadwood, S. Dak.

Cash, Joseph H. A History of Lead, South Dakota, 1876–1900. Unpublished M.A. thesis, Univ. S. Dak., 1959.

Hall, W. F. Education in the Black Hills before 1900. Unpublished M.A. thesis, Univ. S. Dak., 1949.

Journal #1. Hearst Free Library, 1894. The journal is deposited at the Homestake Library, Lead, S. Dak.

Lead City Miners' Union Minute Book, 1890–1901. This book belongs to a private collection, Lead, S. Dak.

Ledger #200. Hearst Free Library, 1897. This is in the possession of the Homestake Library, Lead, S. Dak.

Manuel, Moses. This unpublished and untitled manuscript may be seen at the Homestake Library, Lead, S. Dak.

Minutes of the Board of Education, 1895–1904. These records are in the office of the Lead Independent School District No. 6, Lead, S. Dak.

Myron, Emma George. The History of the Homestake Mine. Unpublished M.A. thesis, Univ. S. Dak., 1928.

DOCUMENTS

Course of Study of the Public Schools of Lead, South Dakota. Lead Public Schools, 1897.

Directory. Lead Public Schools, 1930–31.

Fourth Biennial Report of the Superintendent of Public Instruction. Pierre, S. Dak., 1898.

Official Proceedings of the Eighteenth Annual Convention, Western Federation of Miners. 1910.

Official Proceedings of the Nineteenth Annual Convention, Western Federation of Miners. 1911.

Official Proceedings of the Twentieth Annual Convention, Western Federation of Miners. 1912.

Official Proceedings of the Twenty-First Annual Convention, Western Federation of Miners. 1913.

United States:

U.S. Commission on Industrial Relations. *Final Reports and Testi-*

133

mony on Industrial Relations. Vol. IV, 64th Congress, 1st session, Senate Document n. 22.

Jenney, W. P., and Newton, Henry. *Report on the Resources of the Black Hills of Dakota.* Washington, 1880.

Richardson, James D. *A Compilation of the Messages and Papers of the Presidents.* Vol. VI. Washington, 1897.

Roberts, George E. *Report of the Director of the Mint upon the Production of Precious Metals in the United States During the Calendar Year 1900.* Washington, 1901.

Tenth Census: 1880.

Twelfth Census: 1900.

HOMESTAKE MINING COMPANY RECORDS

Annual Reports, 1878–1942.

Bjorge, Guy N. History of the Homestake Employees Aid Fund, a letter to all members. June 13, 1947.

Employees Record of Underground Service. May 31, 1936.

Foreign-Born. A list of workers. Undated.

Hiring Report, 1939, Employment Department.

Mitchell, D. P. Letter to Emma George Myron. Oct. 21, 1931.

Report of the Hospital Department, 1900–1942.

Report on Silicosis. Sept. 5, 1936.

The Homestake Safety Bulletin. July and Aug. 1918.

The Homestake Story. Lead, S. Dak., 1960.

The Story of Homestake. Lead, S. Dak., 1954.

Yates, B. C. Letter to Homestake employees. Apr. 1, 1919.

PERSONAL INTERVIEWS

Batinovich, Jerry. Oct. 6, 1958.
Clark, Mrs. Horace S. Dec. 10, 1958.
Coleman, Blanche. May 3, 1961.
Dunmire, Joe. Jan. 2, 1964.
Furze, Richard. Dec. 16, 1958.
Howe, Donald P. Jan. 2, 1965.
Hunkins, R. V. Aug. 13, 1958.
Jobe, J. A. Oct. 8, 1958, and May 26, 1961.
Kellar, Kenneth. Apr. 18, 1961, and Dec. 29, 1964.
Kravig, Clarence. Jan. 5, 1965.
McLaughlin, Donald. June 4, 1970.
Mastrovich, John. Mar. 18, 1959.
Morcom, Jelbert. Aug. 13, 1958.
Morganti, Helen. Sept. 4, 1958.
Price, Marjorie Yates. Sept. 1, 1958, and Oct. 12, 1960.
Price, S. Goodale. Apr. 13 and 15, 1961.
Semones, Arthur M. Jan. 7, 1965.
Stabio, Margaret. Sept. 22, 1958.

SECONDARY SOURCES

BOOKS

Andreas, A. T. *Historical Atlas of Dakota.* Chicago, 1884.
Arnot, R. Page. *The Miners.* London, 1949.
Baldwin, George P. (ed.). *The Black Hills Illustrated.* Deadwood, S. Dak., 1904.

Blackburn, William Maxwell. "Historical Sketch of North and South Dakota, 1893." *South Dakota Historical Collections,* I (1902). Aberdeen, S. Dak.

Bonfils, Winifred Black. *The Life and Personality of Phoebe Apperson Hearst.* San Francisco, 1928.

Brissenden, Paul F. *The I.W.W., A Study of American Syndicalism.* New York, 1920.

Brown, Jesse, and Willard, A. M. *Black Hills Trails.* Rapid City, S. Dak., 1924.

Carlson, Oliver, and Bates, Ernest Sutherland. *Hearst, Lord of San Simeon.* New York, 1936.

Casey, Robert J. *The Black Hills and Their Incredible Characters.* New York, 1949.

Collins, Charles, compiler. *Black Hills History and Directory.* Central City, Dakota Territory, 1878.

Coursey, O. W. *The Beautiful Black Hills.* Mitchell, S. Dak., 1926.

Driscoll, R. E., Sr. *Seventy Years of Banking in the Black Hills.* Rapid City, S. Dak., 1948.

Drucker, Peter F. *The Concept of the Corporation.* New York, 1964.

Esling, Dean Arthur. *My Black Hills Story.* Rapid City, S. Dak., 1969.

Fay, Albert H. *A Glossary of the Mining and Mineral Industry,* Bureau of Mines Bulletin No. 95, Washington, 1920.

Fielder, Mildred. *The Treasure of Homestake Gold.* Aberdeen, S. Dak., 1970.

Goodspeed, Weston Arthur. *The Province and the States.* 2 volumes. Madison, Wis., 1904.

Greever, William S. *The Bonanza West.* Norman, Okla., 1963.

Haywood, William D. *Bill Haywood's Book.* New York, 1929.

Hebert, Frank. *Forty Years of Mining and Prospecting in the Black Hills.* Rapid City, S. Dak., 1921.

Howe, Donald P. "Lead." In *Lawrence County,* edited by Mildred Fielder. Lead, S. Dak., 1960.

Howe, F. S. *Deadwood Doctor.* Deadwood, S. Dak., 1951.

Hughes, Richard B. *Pioneer Years in the Black Hills.* Agnes Wright Spring (ed.). Glendale, Calif., 1957.

Hungerford, Edward. *Wells Fargo, Advancing the American Frontier.* New York, 1949.

Hunkins, R. V. "America's Greatest Gold Mine—The Homestake." In *The Black Hills,* edited by Roderick Peattie. New York, 1952.

Hunkins, R. V. "The Black Hills—A Storehouse of Mineral Treasure." In *The Black Hills,* edited by Roderick Peattie. New York, 1952.

Ingalls, Walter Renton (ed.). *The Mineral Industry During 1909,* XVIII: 296. New York, 1910.

Jackson, Donald. *Custer's Gold: The United States Cavalry Expedition of 1874.* New Haven and London, 1966.

Jensen, Vernon H. *Heritage of Conflict.* Ithaca, N.Y., 1950.

Johnson, Willis E. *South Dakota, A Republic of Friends.* Pierre, S. Dak., 1911.

Kingsbury, G. W. *History of Dakota Territory.* 3 volumes. Chicago, 1915.

Kirkland, Edward C. *A History of American Economic Life.* New York, 1947.

Kravig, Clarence N. "Mining in Lawrence County." In *Lawrence County,* edited by Mildred Fielder. Lead, S. Dak., 1960.

Lamar, Howard Roberts. *Dakota Territory 1861–1889.* New Haven, Conn., 1956

MaGuire, Horatio N. *The Coming Empire, 1878.* Quoted in Coursey, O. W., *The Beautiful Black Hills.* Mitchell, S. Dak., 1926.

McClintock, John S. *Pioneer Days in the Black Hills.* Deadwood, S. Dak., 1939.

Meddaugh, J. E. *Souvenir of Lead—Black Hills Metropolis.* Lead, S. Dak., 1892.

Nelson, R. Bruce. *Land of the Dacotahs.* Minneapolis, 1946.

Parker, Watson. *Gold in the Black Hills.* Norman, Okla., 1966.

Paul, Rodman Wilson. *California Gold.* Lincoln, Nebr., 1947.

Paul, Rodman Wilson. *Mining Frontiers of the Far West, 1848–1880.* New York, 1963.

Phillips, Walter Shelley. *The Old-Timer's Tale.* Chicago, 1929.

Quiett, Glenn Chesney. *Pay Dirt.* New York, 1936.

Reed, Jewett V., and Harcourt, A. K. *The Essentials of Occupational Diseases.* Springfield, Ill., 1941.

Rhoads, William. *Recollections of Dakota Territory.* Fort Pierre, S. Dak., 1931.

Rickard, T. A. *A History of American Mining.* New York, 1932.

Rickard, T. A. *The Stamp Milling of Gold Ores.* New York, 1897.

Robinson, Doane. "Black Hills Bygones." *South Dakota Historical Collections,* XII:198–207 (1924).

Robinson, Doane. *A Brief History of South Dakota.* New York, 1905.

Rosen, Peter. *Pa-ha-sa-pah, or The Black Hills of South Dakota.* St. Louis, 1895.

Sanford, Rev. John I. *The Black Hills Souvenir.* Denver, Colo., 1902.

Schell, Herbert S. *History of South Dakota.* Lincoln, Nebr., 1961.

Spring, Agnes Wright. *The Cheyenne and Black Hills Stage and Express Routes.* Glendale, Calif., 1949.

Stokes, George W., and Driggs, Howard E. *Deadwood Gold.* New York, 1926.

Swanberg, W. A. *Citizen Hearst.* New York, 1963.

Tallent, A. D. *The Black Hills: or The Last Hunting Grounds of the Dakotahs.* St. Louis, 1899.

Wolle, Muriel Sibell. *The Bonanza Trail.* Bloomington, Ind., 1953.

Yates, B. C. "Some Features of Mining Operations in the Homestake Mine." *Black Hills Illustrated.* Deadwood, S. Dak., 1904.

ARTICLES

Benfer, W. C. "The Story of the Homestake Lockout." *International Socialist Review,* X:782–90 (Mar. 1910).

Blackstone, R. "The Homestake Mine." *Pahasapa Quarterly,* V:16–30 (June 1916).

Briggs, Harold E. "The Settlement and Economic Development of the Territory of Dakota." *South Dakota Historical Review,* I:151–66 (Apr. 1936).

Brooks, C. A. "Accident Prevention at the Homestake Mine." *Explosives Engineer,* IV:209–12.

"Butte and Lead Contrasted." *Engineering and Mining Journal,* XCVIII: 378 (Aug. 29, 1914).

Fulton, Charles E. "The Cyanide Process in the Black Hills of South Dakota." *South Dakota School of Mines Bulletin,* V:67–68 (1902).

Irwin, E. F. "Homestake Mine—Example in Industrial Cooperation." *Mining Congress Journal,* X:79–80 (1924).

"Lead—August 8." *Engineering and Mining Journal,* XCVIII:324 (Aug. 15, 1914).
Lincoln, Francis Church. "Half a Century of Mining in the Black Hills." *Engineering and Mining Journal,* CXXII:205–14 (Aug. 7, 1926).
"Lucky Homestake." *Fortune* (June 1934).
McClure. "Dakota." *Harpers New Monthly Magazine,* XVIII:347–64 (Jan. 1889).
Moyer, Charles H. "Letter to the Editor." *Miners' Magazine,* pp. 45–46 (Jan. 1902).
O'Harra, C. C. "Custer's Black Hills Expedition of 1874." *Black Hills Engineer,* XVII:228 (Nov. 1929).
O'Harra, C. C. "Early Placer Mining in the Black Hills." *Black Hills Engineer,* XIX:343–61 (Nov. 1931).
Palais, Hyman. "A Survey of Early Black Hills History." *Black Hills Engineer,* XXVII:3–101 (Jan. 1941).
Raddick, W. P. "Early History of Lead." *Lead Daily Call,* Aug. 5, 1926, pp. 1–2.
Ross, A. J. M., and Wayland, R. G. "A Brief Outline of Homestake Mining Methods." *Black Hills Engineer,* XIV:145 (May 1926).
Tracy, William F. "Letter to the Editor." *Lead Daily Call,* Nov. 15, 1909.
"A Trip to the Black Hills." *Scribner's Monthly,* XIII:748–56 (Apr. 1877).
Yates, B. C. "The Homestake Mine." *Black Hills Engineer,* XIV:131–33 (June 1926).
Yates, B. C. "South Dakota Gold Mining." *Engineering and Mining Journal,* CX:1169 (Dec. 18, 1920).
Yates, B. C. "Welfare Work at the Homestake Mine." *Engineering and Mining Journal,* CX:198–203 (July 31, 1920).
"Work Suspended in All Departments." *Deadwood Pioneer Times,* Nov. 25, 1909.

PERIODICALS

Black Hills Pioneer, 1876–1900.
Deadwood Pioneer Times, 1900–1965.
Engineering and Mining Journal, 1905–40.
Evening Call, Lead, 1896–1908.
Illustrated Booster Nugget, Lead High School, Lead, S. Dak., 1912.
Lantern, 1907–10.
Lead City Daily Tribune, 1881–1900.
Lead Daily Call, 1900–1972.
Miners' Magazine, 1902–10.
Sharp Bits, 1949–65.
Telegraph-Herald, Lead City and Central City, 1878.

INDEX